ENERGY 2016

The Sixth International Conference on Smart Grids, Green Communications and IT Energy-aware Technologies

June 26 - 30, 2016

Lisbon, Portugal

ENERGY 2016 Editors

Steffen Fries, Siemens AG, Germany

Claus-Peter Rückemann, Leibniz Universität Hannover / Westfälische Wilhelms-Universität Münster / North-German Supercomputing Alliance (HLRN), Germany

Printed from e-media with permission by:

Curran Associates, Inc.
57 Morehouse Lane
Red Hook, NY 12571

Printed by Curran Associates, Inc. (2016)

International Academy, Research, and Industry Association (IARIA)
412 Derby Way
Wilmington, DE 19810

Phone: (408) 893-6407
Fax: (408) 527-6351

petre@iaria.org

Additional copies of this publication are available from:

Curran Associates, Inc.
57 Morehouse Lane
Red Hook, NY 12571 USA
Phone: 845-758-0400
Fax: 845-758-2633
Email: curran@proceedings.com
Web: www.proceedings.com

Table of Contents

Energy Saving Potential of Adaptive, Networked, Embedded Systems - A Case Study 1
Patrick Heinrich, Erik Oswald, and Rudi Knorr

The TUCool Project - Low-cost, Energy-efficient Cooling for Conventional Data Centres 10
Matthias Vodel and Marc Ritter

Ensuring Secure Communication in Critical Infrastructures 15
Steffen Fries and Rainer Falk

Aggregators Efficiency in Distributed Power Networks 21
Alain Tcheukam Siwe and Hamidou Tembine

Optimization of Power Usage Effectiveness for Heterogenous Modular Data Centers using Neural Network 27
Vishal Kumar Singh and Jinhua Guo

SmartEnergyHub - A Big-data Approach for the Optimization of Energy-intensive Infrastructures 33
Florian Maier, Andreas Wohlfrom, Falko Koetter, Sabrina Merkel-Malkowski, and Daniel Zech

Study of a Smart Energy Community PV and Storage Requirements. A Modeling Approach Towards Net-Zero Energy. 37
Emilio J. Palacios-Garcia, Antonio Moreno-Munoz, Isabel Santiago, Isabel M. Moreno-Garcia, and Rafael J. Real-Calvo

Finding Needles in a Haystack: Line Event Detection on Smart Grid PMU Data Streams 42
Duc Nguyen, Scott Wallace, and Xinghui Zhao

Transforming Utilities: Turning Data into Intelligence with Data Analytics 48
Vanesa Cackovic and Zeljko Popovic

Multi-objective Optimization of Energy Hubs at the Crossroad of Three Energy Distribution Networks 54
Giuseppe Paterno, Diego Arnone, Alessandro Rossi, Massimo Bertoncini, Eleonora Riva Sanseverino, and Mariano Giuseppe Ippolito

ENERGY 2016

Foreword

The Sixth International Conference on Smart Grids, Green Communications and IT Energy-aware Technologies (ENERGY 2016), held between June 26 - 30, 2016 - Lisbon, Portugal, continued the event considering Green approaches for Smart Grids and IT-aware technologies. It addressed fundamentals, technologies, hardware and software needed support, and applications and challenges.

There is a perceived need for a fundamental transformation in IP communications, energy-aware technologies and the way all energy sources are integrated. This is accelerated by the complexity of smart devices, the need for special interfaces for an easy and remote access, and the new achievements in energy production. Smart Grid technologies promote ways to enhance efficiency and reliability of the electric grid, while addressing increasing demand and incorporating more renewable and distributed electricity generation. The adoption of data centers, penetration of new energy resources, large dissemination of smart sensing and control devices, including smart home, and new vehicular energy approaches demand a new position for distributed communications, energy storage, and integration of various sources of energy.

We take here the opportunity to warmly thank all the members of the ENERGY 2016 Technical Program Committee, as well as the numerous reviewers. The creation of such a high quality conference program would not have been possible without their involvement. We also kindly thank all the authors who dedicated much of their time and efforts to contribute to ENERGY 2016. We truly believe that, thanks to all these efforts, the final conference program consisted of top quality contributions.

Also, this event could not have been a reality without the support of many individuals, organizations, and sponsors. We are grateful to the members of the ENERGY 2016 organizing committee for their help in handling the logistics and for their work to make this professional meeting a success.

We hope that ENERGY 2016 was a successful international forum for the exchange of ideas and results between academia and industry and for the promotion of progress in the fields of smart grids, green communications and IT energy-aware technologies.

We are convinced that the participants found the event useful and communications very open. We also hope that Lisbon provided a pleasant environment during the conference and everyone saved some time for exploring this beautiful city.

ENERGY 2016 Chairs:

ENERGY Advisory Chairs

Stefan Mozar, CCM Consulting / CQ University - Sydney International Centre, Australia
Mardavij Roozbehani, Massachusetts Institute of Technology, USA
Mark Apperley, University of Waikato, New Zealand
Steffen Fries, Siemens, Germany

ENERGY Industry Liaison Chairs
Marco Di Girolamo, Hewlett-Packard Company - Cernusco sul Naviglio, Italy
Dragan Obradovic, Siemens AG, Germany

ENERGY Special Area Chairs on Nano-Grids
Peter Müller, IBM-Zurich, Switzeland

ENERGY Special Area Chairs on Smart Grids
Ritwik Majumder, ABB AB / Corporate Research Center - Vasteras, Sweden

ENERGY Special Area Chairs on IT-energy- aware, Planning
Brian P. Gaucher, IBM Research Division - Yorktown Heights, USA

ENERGY Special Area Chairs on Grid, Green Communication
Gargi Bag, ABB Corporate Research, Sweden
Ken Christensen, University of South Florida, USA

ENERGY Special Area Chairs on Vehicular
Grzegorz Swirszcz, IBM Watson Laboratory, USA

ENERGY 2016

Committee

ENERGY Advisory Committee

Stefan Mozar, CCM Consulting / CQ University - Sydney International Centre, Australia
Mardavij Roozbehani, Massachusetts Institute of Technology, USA
Mark Apperley, University of Waikato, New Zealand
Steffen Fries, Siemens, Germany

ENERGY Industry Liaison Chairs

Marco Di Girolamo, Hewlett-Packard Company - Cernusco sul Naviglio, Italy
Dragan Obradovic, Siemens AG, Germany

ENERGY Special Area Chairs on Nano-Grids

Peter Müller, IBM-Zurich, Switzeland

ENERGY Special Area Chairs on Smart Grids

Ritwik Majumder, ABB AB / Corporate Research Center - Vasteras, Sweden

ENERGY Special Area Chairs on IT-energy- aware, Planning

Brian P. Gaucher, IBM Research Division - Yorktown Heights, USA

ENERGY Special Area Chairs on Grid, Green Communication

Gargi Bag, ABB Corporate Research, Sweden
Ken Christensen, University of South Florida, USA

ENERGY Special Area Chairs on Vehicular

Grzegorz Swirszcz, IBM Watson Laboratory, USA

ENERGY 2016 Technical Program Committee

Amir Abtahi, Florida Atlantic University - Boca Raton, US
Nizar Al-Holou, University of Detroit Mercy, USA
Ahmed Al-Salaymeh, University of Jordan, Jordan
Lachlan Andrew, Swinburne University of Technology – Melbourne, Australia
Luca Ardito, Politecnico di Torino, Italy
Diego Arnone, Engineering Ingegneria Informatica S.p.A., Italy
Mehdi Bahrami, University of California, Merced, USA

Copyright Information

Energy Saving Potential of Adaptive, Networked, Embedded Systems - A Case Study

Patrick Heinrich[1], Erik Oswald[1] and Rudi Knorr[1,2]

[1]Fraunhofer Institute for Embedded Systems and Communication Technologies ESK, Munich, Germany
[2]Chair for Communication Systems, Institute of Computer Science, University of Augsburg, Germany
E-Mail: forename.surname@esk.fraunhofer.de

Abstract—**This paper presents and evaluates the energy saving potential of adaptive, networked, embedded systems. The aim is to demonstrate the benefits of modeling the energy demand during the development of such systems. For this purpose, the previous developed energy model is applied within a case study and different allocations of software components are compared. The estimated energy demands of these allocations are presented and discussed. The analyzed system of the case study represents an automotive system which executes two advanced driver assistance applications. The system is adaptive, which means that temporally unnecessary applications will be deactivated. Within the evaluated system this deactivation depends on the vehicle speed, which is derived by the New European Driving Cycle. Two different allocations of software components are evaluated. One represents the today's allocation, the other an energy-focused allocation. The case study shows an energy saving potential of about 18 %.**

Keywords–Embedded systems; Adaptivity; Networked systems; Energy estimation; Automotive; Case study.

I. Introduction

Networked, embedded systems of modern cars consist of up to 80 independent electronic control units (ECUs), which cooperate to enable complex applications. These systems are characterized by a high degree of interaction and consist of different (specialized) communication networks [1]. Future cars will probably enable autonomous driving, which further increases the number of ECUs and the interaction between these ECUs. However, future automobiles will also be adaptive to decrease the energy demand by deactivating temporarily unnecessary applications. Hereby, the chosen allocation of software components on the network's ECUs is relevant w.r.t. the energy demand [2]. This is caused by different applicable energy saving states and energy demand necessary to (de)activate hardware and software. The allocation is done by system designers relatively early within the development process [3], which makes it necessary to estimate the later energy consumption at that time using the available information.

In today's luxury-class vehicles for instance, the electrical and electronic components draw up to 2.5 kW [4][5]. Compared to what the vehicle engine requires (e.g., 55 kW), 2.5 kW seems small. However, the electrical components consume energy during every mode of operation, even when in standby mode. The vehicle engine consumes most of its energy during acceleration and even here the maximum power is seldom demanded. Communication and sensors/actuators cause an increasing amount of energy consumption of networked embedded systems. An increase of 100 Watt thus means that fuel consumption rises by 0.1 liter per 100 km, leading to an increase in CO_2 emissions of 2.5 gram per km [4]. This illustrates the considerable relevance for energy savings, an aspect that should be factored in during the development.

This paper presents an automotive case study to demonstrate the energy saving potential, which is usable when the energy demand of such systems is modeled previously to the allocation of software components. In Section II, the previous work and in Section III the characteristics of the system are presented on which the case study is based. The case study is described within Section IV. In Section V and VI, the energy demand estimation is presented and discussed. Finally, Section VII concludes the paper.

II. Related Work

The energy models used to estimate the energy demand within this case study are based on previous work, which has focused the various elements of adaptive, networked, embedded systems. The necessity to model the energy demand of adaptivity and the uncommon fact that suboptimal (w.r.t the energy demand) allocations of software components per system state may form an optimized adaptive system was described within [6]. Different approaches to estimate the energy demand previous to system integration, e.g., during development of the system, were presented within [7]. Based on this work the further developed models focused on the energy demand estimation during the development phase where the software components are allocated within the networked system. This enables a good trade-off between the inaccuracy of early energy estimations and the available energy saving potential of the later system – which is reduced after every decision within the development process [7]. Within [8] a model is presented which enables the energy demand estimation of software components executed on embedded systems. The estimation is based on program flowcharts and enables an accuracy between -11.9 % and +6.9 %. Embedded systems (ECUs) including sensors/actuators and offset energy demand were modeled within [9]. This also includes the effects of parallel execution of software components on the energy demand of adaptive ECUs. This work focused on aspects of adaptive systems where temporarily unnecessary software components are deactivated. The energy demand of communication between embedded system is modeled within [10]. These communication networks

$$E_{total} = \underbrace{\underbrace{\sum_{i=1}^{\#ECUs} \left[\overbrace{\sum_{j=1}^{\#SWCs} (E_{flow_j} \cdot \tau_{corr})}^{\text{software component}} + E_{SensAkt_i} + E_{ECU_i} \right]}_{\text{adaptive, embedded system}} + \overbrace{E_{com}}^{\text{communication}}}_{\text{adaptive, networked, embedded system}} \qquad (1)$$

are characterized by heterogeneity and a high degree of interaction between the specialized parts of the network. The presented model is based on individual communication connections which significantly simplifies the energy demand estimation of adaptive systems. The model enables estimations within an error range of -2.4 % and +2.8 %, which is very accurate.

Equation (1) summarized the presented parts of the energy model. The energy demand of the system is represented by E_{total}. The networked system consist of a number of ECUs ($\#ECUs$) which communicate via communication networks demanding the energy amount represented by E_{com}. The energy demand to execute software components on an ECU is represented by the number of software components ($\#SWCs$), the energy demands of the program flowcharts (E_{flow}) and the correction factor τ_{corr} (cf. [8]). The symbol $E_{SensAkt}$ represents the energy demand of sensors and actuators, the symbol E_{ECU} the offset energy demand of ECUs. The detailed calculation of the symbols is presented within the considered previous work.

In the following section, the characteristics of adaptive, networked, embedded systems are presented.

III. Adaptive, Networked, Embedded Systems

Networked, embedded systems as found within the automotive sector are characterized by a high degree of heterogeneity [11]. This means different kind of ECUs by various suppliers and different types of communication protocols and topologies resulted by specialized technologies. A common hardware architecture of networked embedded systems is shown within Figure 1. Additionally, these systems are characterized by a high degree of interaction, which means that applications (which are experienced by the user) are spread over various ECUs within the networked, embedded system. These individual components are commonly used for more than one application, i.e., the rain sensors is used to control the windshield wipers and also to parametrize the Electronic Stability Program (ESP). Adaptivity within embedded systems enables the deactivation of temporarily unnecessary functionality including its hardware. This enables the design of more energy efficient systems, but significantly enlarges the number of possible system designs. Furthermore, factors such as time and energy for (de)activation and availability of different kind of sleep modes are relevant now. Adaptivity necessitates considering the whole systems instead of just looking at individual components, because it is

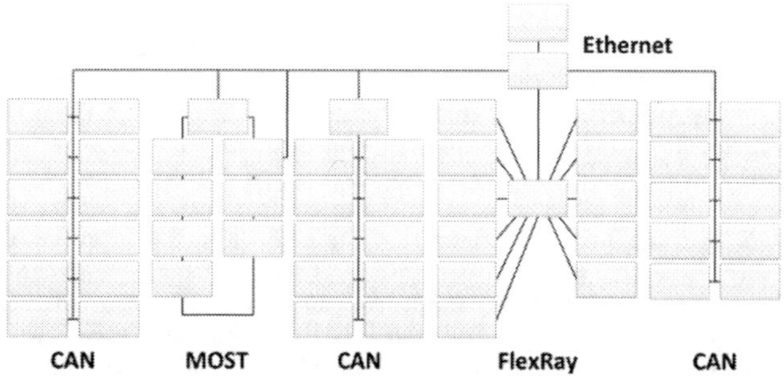

Figure 1. Typical networked embedded system (BMW 7 series) [adapted from BMW 2005, quoted from [12]]

possible that individual suboptimal systems form together a more energy efficient system [6].

Embedded systems within the automobiles commonly execute safety-critical applications, which results in real-time constraints, for example concerning transmission and response times. System designers of networked, embedded systems are also faced with a lot of other constraints and limitations (e.g. performance and memory limitations, timing constraints and heterogeneity). The energy demand is just one of the aspects system designer have to consider. However, this aspect has become increasingly important over the last few years.

In the following section, details concerning hardware and software architecture and the adaptivity of the system considered within the case study are presented.

IV. Case Study

Within this section the advanced driver assistance systems considered in the case study are explained and the resulted software architectures are presented. Afterwards, the considered hardware architecture including the energy relevant parameters are discussed. And finally the adaptivity of the system and the relevant system context is pointed out.

A. Advanced Driver Assistance Systems

The two advanced driver assistance systems which software components are allocated within the networked system are Adaptive Cruise Control (ACC) and Automatic Parking Assist (APA), which are explained in the following.

1) Adaptive Cruise Control: This system automatically adjusts the vehicle speed to maintain a given distance from a vehicle ahead. The distance is measured using sensors like radar sensors. To enable such a functionality different electronic control units (ECUs) have to interact, e.g., within an AUDI A8 80 ECUs are part of this system [13]. An ACC system consist of different components (cf. [14][13][15] [16][17]). An object recognition is used to detect vehicles ahead and determine the absolute speed. Therefore long-range and short-range radar sensors are used. Additionally, the data is used to predict the probable course of the own vehicle to avoid the detection of irrelevant objects. To do so, additional input from a path and curve calculation is necessary, which uses data from yaw rate and steering angle sensors. The main components of ACC are the control loops to control the vehicle speed and to follow the vehicle ahead. Depending on the final implementation, the calculated required torque or deceleration is transmitted to engine or brake control. Figure 2 represents the derived software architecture of the ACC system as it is used within the case study. A long-range and a short-range radar including the signal processing are used to recognize objects (vehicles) ahead. Sensors for wheel speed, yaw rate and steering angle are used to analyze the current and future status of the own vehicle (including path and curve calculation and course prediction). This information is used to enable the control loops, which are summarized within "ACC Control". Engine and brake control execute the calculated acceleration or deceleration. The arrows – refined with the number of bytes transfered – represent communication.

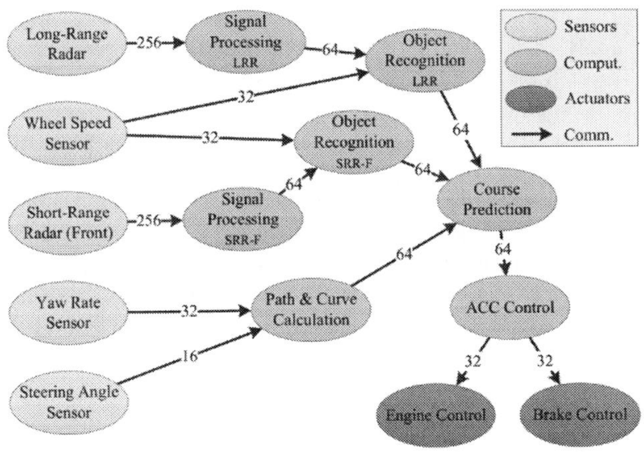

Figure 2. Software architecture of Adaptive Cruise Control

2) Automatic Parking Assist: This driver assistance system enables a driver to automatically park into suitable parking spaces (cf. [18]). Short-range radar or ultrasonic sensor at front and rear are used to measure the distance between the vehicles. The system controls engine, brake and steering wheel to maneuver and park the vehicle automatically. Sensors for wheel speed, yaw rate and steering angle are used to monitor the vehicle's parking operations. Figure 3 shows the derived software architecture of the automatic parking assist system. Two radar sensors including signal processing and object recognition observe front and rear of the vehicle. Wheel speed, yaw rate and steering angle sensor are additionally used to control the parking assistant. Engine, brake and steering control execute the calculated acceleration, deceleration and steering.

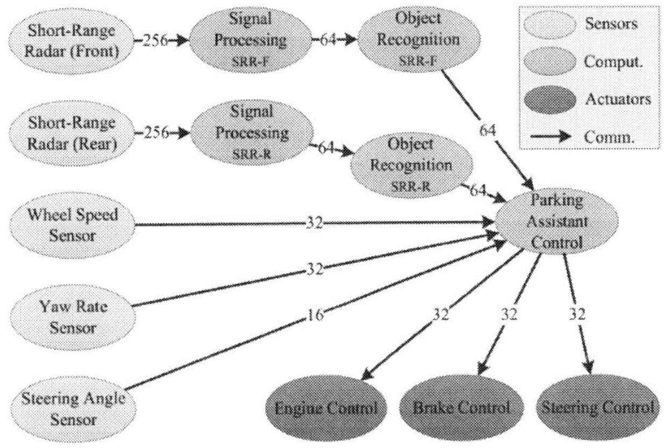

Figure 3. Software architecture of Automatic Parking Assist

Within this case study the driver assistance systems Adaptive Cruise Control and Automatic Parking Assist are used to represent an automotive system. These systems are real-time critical, which results in a suitable high control frequency. Hansson et al. [19] define the control frequency of ACC with 10 Hz, which represents an response time of 100 ms. This value is used as cycle time of the ACC's and APA's software components. Furthermore, some of the software components are represented within both applications. If ACC and APA are active at the same time, the identical software components are executed just once and provides the required information to the following software components of both applications. Table I presents the relevant parameters of the software components, which i.a. are derived from [19] and [20]. The relevant parameters are: execution time, cycle, time, deadline and produced (data) output.

In the following subsection, the hardware architecture and the energy relevant parameters of the system are introduced.

B. Hardware Architecture and Energy Parameters

Within this subsection, the properties of the system are presented. That means in detail, ECUs including sensors and actuators, software components and adaptivity. The hardware architecture of the case study is based on the networked, embedded system of a BMW 7 series (as shown in Figure 1). The chosen hardware architecture is presented within Figure 4, where the topology is similar but the number of ECUs is reduced. The case study's validity is not influenced through that reduction, because grouping software components is focused by both considered "allocation styles" (cf. Subsection VI-A). Equal distribution of software components normally increases the energy demand, caused by the offset energy demand of ECUs, which is independent of the number of software components running on an ECU.

Within the case study just one ECU is defined, which is equipable with different sensors and actuators. The reason for this is that ECUs with different energy efficiencies would concentrate the software components on the most energy

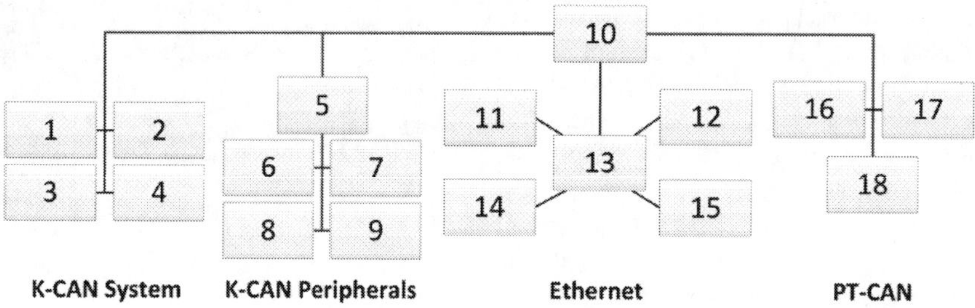

Figure 4. Simplified network topology of the case study, based on the networked, embedded system of the BMW 7 series (cf. Figure 1). Note: The numbers are used to identify the ECUs during allocation of the software components.

efficient ECUs. However, the purpose of this case study is to demonstrate the energy saving potential of adaptive, networked, embedded systems. Different ECUs would increase complexity without an advantage. Therefore, just one energy saving state of the components is defined. E.g. the µController has three states – running, idle and sleep. The characteristics of the used µController are derived from existing µControllers like the "MPC5674F" [21] of Freescale Semiconductor, which works at 264 MHz and has a power demand of 850 mA at 1.32 Volt. A current consumption of 100 mA during idle operations is assumed. The activation time depends on the phase-locked loop (PLL) lock time and was specified with 400 µs. The time for deactivation is assumed to bei 50 µs. This results in an energy demand of 0.5 mWs per deactivation including activation of the µController. The offset power demand of the defined ECU is specified with 1.5 W, which includes the energy demand of components such as the power supply unit and the voltage regulators. Weber et al. [22] have defined a Body Control Module with a power consumption of 10.75 W, where 1 W is used for the µController and 4 W is defined as ECU offset power consumption. The ECU offset power consumption within the case study is defined at the lower range of possible values, because the full capacity of the defined system is not used by the two driver assistant systems. Through that a falsification of the percentage ration of the energy demand within the system is prevented. The effect on other components is low, because most scale with the utilization of the ECU. (This is not the case for the states idle or sleep. However, the energy demand is lower and through that estimations errors have a lower impact.)

Due to works which try to reduce the energy demand of sensors and actuators (such as Benbasat [23]) the activation time is assumed to be much smaller than the cycle time of the used sensors and actuators. That means these components are deactivateable after the software component execution. (This is not the case for short-term interruption caused by scheduling.) This is another reason why activation time of sensors and actuators are necessary to reach more accurate energy demand estimations. Radar systems within the automotive area have a power demand of about 4 to 4.5 W (cf. [24] and [25]). The long-range radar used within this case study is defined to have a power consumption of 4.5 W and an activation time of 25 ms. The short-range radar consumes 1.75 W of power and need 25 ms to get active. The power consumption of "Engine-", "Brake-" und "Steering-Control" are derived from the Automotive Engine Control IC [26] of Freescale Semiconductors. (Note: Components such as relays, servomotors, spark plugs, etc. are not concerned within this work. That is why just the control unit is considered.) The Engine Control IC has a voltage range of 4.5 to 36 V with a maximum current of 14 mA. The power demand is defined to be 168 mW (12 V, 14 mA) and the activation time to be 1 ms. Rotations speed sensors such as "KMI17/4" [27] of NXP Semiconductors are defined with a turn on delay time of 1 ms. (This value is also used for the wheel speed and the angle sensor.) The energy demand is defined with 120 mWs, because a voltage of 12 V is assumed, which results a current of 10 mA. Angle sensors such as "KMA220" [28] of NXP Semiconductors need a voltage supply of 5 V and have a current demand of 10 to 21 mA. Through that, the "Steering Angle Sensor" of this case study is assumed to have a power demand of 75 mW. The properties of the used yaw rate sensor is derived by the combined inertial sensor for vehicle dynamics control "SMI650" [29] of Robert Bosch GmbH. The data sheet specifies a necessary voltage supply of 5 V and a current demand of 25 mA. This results in a power demand of 125 mW. An activation time of 650 ms is specified, which includes a detailed self test. It is assumed that the self test is just relevant during the initial start of the system and not after every sleep state. Through that an activation time of 5 ms is defined for the case study. Table II shows the derived parameters concerning the energy demand of sensors and actuators including the activation time. (Note: This work focuses on the electronic system, e.g., the components like the engine, which is controlled by "Engine Control" is excluded.)

The energy relevant parameters of the communication networks are derived from the measured values at [10]. The parameters for Ethernet are derived from data sheets. The CAN network within the case study uses a data rate of 500 kbit/s. Due to the specification [30] eight bytes per message are user data and 44 bits are communication overhead. The energy relevant parameters of the "High-Speed CAN Transceiver MCP2561" [31] and of the "Stand-Alone CAN Controller with SPI Interface MCP2515" [32] are used. The Ethernet network works with a data rate of 10 Mbit/s. A maximum of 1500 bytes per message are user data and 144 bits are communication overhead. The Ethernet transceiver "LAN8710A" [33] and the Ethernet controller "CS8900A" [34] were served as basis to derive the energy relevant parameters. However, a more efficient sleep state was assumed. Table III summarizes the energy relevant values of the used communication networks.

Within the work which models the energy demand of embedded communication networks [10] a energy demand to transfer data between networks is defined as "transfer energy". This energy demand depends on the used µController and the network on both sides of the gateway. Identical networks result a smaller energy demand, than different networks, e.g., because of the need to split user data or handle different data rates. Within the case study a energy demand of 0.12 mWs per message is defined and a transmission time of 0.1 ms. If the gateway is a specialized router or switch, than half of the energy demand is assumed. Caused by unproven values, the energy parameters are estimated at the lower range of the possible values to prevent distortions.

The energy demand necessary for the system's adaptivity is determined by the (de)activation of software components and ECUs. The energy demand to activate or deactivate a software component is determined to be 5.61 mWs. This value is estimated by the summarized value of 5 ms for self test, safe/load data, etc. and the power consumption of the used µController of 1122 mW. The (de)activation of entire ECUs is defined with the energy demand of 2.5 Ws. This value is resulted from the power consumption of a Body Control Modul [22], which has a power demand of 10 W and an activation time of 250 ms. An activation time of 100 to 200 ms is necessary to boot up the ECU and the additionally need to receive all necessary messages to get the actual data to work with [35]. This results in the value of 2.5 Ws.

Within the following subsection the characteristics of the system concerning adaptivity are presented.

C. Adaptivity

The analyzed system within this case study is adaptive as presented within Section III, i.e., deactivates temporarily unnecessary hardware and software components. (The necessary energy demand was already presented within the subsection before.) The need of a specific application arises by an user request or the actual context (environment) of the vehicle. For example, windshield wipers are just necessary, if there is rain or other kind of water at the windshield, or an automatic parking assist is not necessary at high speed.

The analyzed driver assistant systems within this case study are activated depending on the current speed of the vehicle, i.e., the relevant system context is "Vehicle Speed". This input is relevant to analyze the energy demand of the system. Within this case study the very common New European Driving Cycle (NEDC) is used to derive the speed profile. The NEDC is used to measure a compareable fuel consumption of vehicles by the manufacturer. The NEDC is standardized within the "Directive 98/69/EC of the European Parliament and of the Council" [36] and defines a drive of 11 kilometers within 1180 seconds. Four identical phases represent urban driving (average speed: 19 km/h) and a phase which represents an overland tour (maximum speed: 120 km/h). Figure 5 shows the resulting speed profile over time.

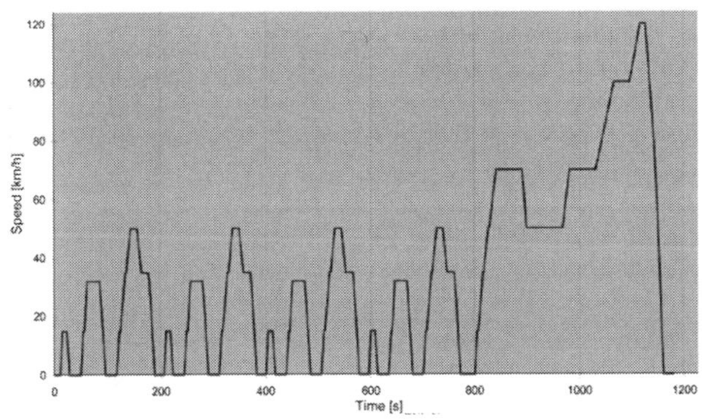

Figure 5. Speed profile over time derived from the New European Driving Cycle (NEDC) [Adapted from [37]]

The speed profile is used to determine the necessity of software components, sensors and actuators during the different speeds of the vehicle. ACCs are commonly activated at a speed larger than 30 km/h. An automatic parking assist can be reasonable used at a speed below 15 km/h. Table I shows the necessary components at three speed ranges (Column 2 to 4), which are necessary to enable the needed functionality of the vehicle.

In the following subsection, the implemented simulation is briefly outlined and the assumptions made are specified.

V. Discrete Event-Driven Simulation

Within this section an event-driven simulation is presented, which is applied to the previously presented system of the case study to estimate the energy demand of two different allocations of software components. The developed simulation uses the concept of an event-driven simulation as presented by Banks et al. [38] (cf. also [39] and [40]) and is implemented as discrete event-driven simulation, because the system only changes after discrete events. The considered system is adaptive (cf. Section III), i.e., the energy demand depends on the system context and the previously executed system state. That means it is necessary to simulate the context of the system as it is defined by Zeigler et al. [41], where the model has to generate the necessary data and the simulation the model's behavior.

The simulation focuses the allocation of software components within a given hardware architecture. This enables to optimize the allocation of software components or the optimized integration of new software components into an existing system. (The usage within Design Space Explorations would be possible, but is not considered within this work.) The position of necessary sensors and actuators can be predetermined, if there are hardware restrictions, or the position is resulted from the allocation of the affiliated software component. A mixture is also possible. However, within this case study the position of sensors and actuators are not predetermined. Furthermore, the

TABLE I. STATES OF SOFTWARE COMPONENTS, SENSORS AND ACTUATORS DURING DIFFERENT VEHICLE SPEEDS AND THEIR SOFTWARE RELEVANT PARAMETERS

Software Component, Sensors, Actuators	$v < 15$ km/h	$v \geq 15$ km/h $v \leq 30$ km/h	$v > 30$ km/h	Exec. Time	Cycle Time	Deadline	Output
Long-Range Radar	OFF	OFF	ON	25 ms	100 ms	100 ms	256 Bit
Signal Processing (LRR)	OFF	OFF	ON	9 ms	100 ms	100 ms	64 Bit
Object Recognition (LRR)	OFF	OFF	ON	30 ms	100 ms	100 ms	64 Bit
Short-Range Radar (Front)	ON	OFF	ON	15 ms	100 ms	100 ms	256 Bit
Signal Processing (SRR-F)	ON	OFF	ON	9 ms	100 ms	100 ms	64 Bit
Object Recognition (SRR-F)	ON	OFF	ON	30 ms	100 ms	100 ms	64 Bit
Short-Range Radar (Rear)	ON	OFF	OFF	15 ms	100 ms	100 ms	256 Bit
Signal Processing (SRR-R)	ON	OFF	OFF	9 ms	100 ms	100 ms	64 Bit
Object Recognition (SRR-R)	ON	OFF	OFF	30 ms	100 ms	100 ms	64 Bit
Path & Curve Calculation	OFF	OFF	ON	25 ms	100 ms	100 ms	64 Bit
Course Prediction	OFF	OFF	ON	25 ms	100 ms	100 ms	64 Bit
ACC Control	OFF	OFF	ON	5 ms	100 ms	100 ms	2x 32 Bit
Parking Assistant Control	ON	OFF	OFF	20 ms	100 ms	100 ms	3x 32 Bit
Engine Control	ON	ON	ON	3 ms	30 ms	30 ms	-
Brake Control	ON	ON	ON	3 ms	30 ms	30 ms	-
Steering Control	ON	ON	ON	3 ms	30 ms	30 ms	-
Wheel Speed Sensor	ON	ON	ON	5 ms	30 ms	30 ms	32 Bit
Steering Angle Sensor	ON	OFF	ON	5 ms	30 ms	30 ms	16 Bit
Yaw Rate Sensor	ON	OFF	ON	10 ms	50 ms	50 ms	32 Bit

TABLE II. ENERGY RELEVANT PARAMETERS OF SENSORS AND ACTUATORS

Sensor/Actuator	Power Consump.	Activation Time
Long-Range Radar	4.500 mW	25 ms
Short-Range Radar	1750 mW	25 ms
Engine Control	168 mW	1 ms
Brake Control	168 mW	1 ms
Steering Control	168 mW	1 ms
Wheel Speed Sensor	120 mW	1 ms
Steering Angle Sensor	75 mW	1 ms
Yaw Rate Sensor	125 mW	5 ms

TABLE III. ENERGY RELEVANT PARAMETERS OF COMMUNICATION

Component	CAN TRX	CAN CC	Eth. TRX	Eth. CC
TX	152 mW	1.1 mW	35.5 mW	8.2 mW
TX#Nodes	13.1 mW	0.02 mW	-	-
RX	2.5 mW	0.3 mW	16.5 mW	2.1 mW
RX#Nodes	0.01 mW	0.01 mW	-	-
Offset	22 mW	23 mW	44 mW	154.7 mW
Sleep	50 µW	20 µW	560 µW	330 µW

allocation of the specific software components are identical within the different system states, i.e., a reallocation of software components during runtime is not considered within this case study. (Simulation and the energy model are able to cope with a reallocation during runtime of the system, but this is not considered within this work.)

Equation (1) is used within the implemented simulation to estimate the energy demand of the later system. Some assumptions which specify the behavior of the system are necessary to made to determine the equation input data.

The execution times of software components vary from ECU to ECU, where the processor frequency has a large influence. In accordance with Walla et al. [42], the execution time is assumed to be inversely proportional to the processor frequency. On processors the software components are scheduled by the Round Robin algorithm, which is preemptive and based on fixed time slices for each process. It is assumed that the time slices are much shorter than the execution times of the software components, i.e., execution times are enlarged, if more than one software component is active at the same time. Furthermore, it is assumed, that software component specific sensors and actuators are only active, when the affiliated software component is active. This results in an energy demand to (de)activate sensors and actuators, however, as described within Subsection IV-B this energy is less than the energy needed to stay active. Further power management techniques like Dynamic Voltage Scaling are not used, because executing software components at maximum processor speed and deactivation afterwards is more energy efficient, if the entire system is considered [43]. Additionally, it is assumed that the communication data rate is used efficiently, i.e., splitting payload is just done, if necessary. Through that the necessary overhead is determinable, even if no further details concerning the kind of communication is available.

To simulate the system context the given speed profile of the NEDC is analyzed to divide the time sequence into slices with non-changing system states. Equation (1) enables to determine the energy demand of every system states. The transitions between system configurations are considered using the necessary energy demands for activation and deactivation of software and hardware components as described within Subsection IV-B.

In addition, different constraints are checked to verify the validity of the allocation. This includes obtaining deadlines of software components, maximum data rate of a network and workload of CPUs. The CPU workload is determined using the resulting workloads of the software components. The workload is not allowed to exceed 100 % or a given maximum value. The data rate is verified using the transmitted data per network, if bus based communication is used. Using full-duplex Ethernet the data rate per interface and direction is analyzed. It is also possible to extend the constraint checks, however, this is not focus of this work. The deadline of software components are examined by comparing the execution time after the scheduling process. Within this case study it is assumed that the deadline of the software components is equal to the cycle time.

In the following section, the examined allocations of software components within the system are detailed and the results of the energy demand estimation are presented and discussed.

VI. Energy Demand Estimation

The energy demand of the presented system (Section IV) is estimated using the introduced simulation (Section V), which is based on the given energy estimation model (Section II). The speed profile provided by the NEDC (cf. Subsection IV-C) is used as system context, which results 48 configuration changes during the NDEC according to the thresholds shown in Table I.

Within the following, the two considered allocations of software components are presented. Afterwards, the estimated energy demand of the system is presented and discussed.

A. Allocation of Software Components

Two possible allocations of software components within the hardware architecture are considered. A so-called "Function-based Allocation", which represents the current usual kind of allocation. That means software components associated to a specific functionality are grouped together on one or just a few hardware components. This is due to the fact that suppliers had previously provided hardware and software components as an integrated system. This paradigm changes nowadays and software components become independent from hardware components, e.g., using standardized software architectures such as AUTOSAR [44]. The other allocation is called "Energy-focused Allocation" where the software components are placed to reduce the energy demand of the system. Within an adaptive system it is relevant to take the usage of the system into consideration, e.g., it is advantageous to group software components which are needed at the same time. This reduces for example the energy demand to activate ECUs. Table IV shows the allocation of the software components to the hardware components of the presented hardware architecture (Section IV-B).

In the following subsection, the estimated energy demands using these allocations are presented.

B. Energy Demand Estimation

Table V shows the resulted energy demand estimation of the two previously presented allocations of software components within the hardware architecture. Beside the energy demand of the entire system, the percentage of energy demand for CPU, ECU offset, sensors/actuators, communication and adaptivity is

TABLE IV. ALLOCATION OF SOFTWARE COMPONENTS, SENSORS AND ACTUATORS ON ECUS (ECU NUMBERS ACCORDING TO FIGURE 4)

Software Component, Sensor, Actuator	Allocation "Func.-based"	Allocation "En.-focused"
Long-Range Radar	#16	#2
Signal Processing (LRR)	#16	#1
Object Recognition (LRR)	#16	#1
Short-Range Radar (Front)	#2	#2
Signal Processing (SRR-F)	#2	#1
Object Recognition (SRR-F)	#4	#1
Short-Range Radar (Rear)	#2	#2
Signal Processing (SRR-R)	#2	#1
Object Recognition (SRR-R)	#4	#1
Path & Curve Calculation	#16	#17
Course Prediction	#16	#17
ACC Control	#16	#17
Parking Assistant Control	#4	#17
Engine Control	#18	#16
Brake Control	#16	#16
Steering Control	#12	#16
Wheel Speed Sensor	#17	#18
Steering Angle Sensor	#12	#18
Yaw Rate Sensor	#17	#18

shown. (This work focuses on the electronic system, e.g., the components like the engine, which is controlled by "Engine Control" is excluded.) Table VI further details the energy demand for communication (cf. [10]) and adaptivity (cf. Subsection IV-B).

In the following subsection, the presented energy demand estimations are discussed.

C. Discussion

As shown within Table V, the energy demand of the "Energy-focused Allocation" is about 18 % less than of the "Function-based Allocation". The column "Difference" shows that the energy for CPU, ECU offset, sensors/actuators and communication is decreased and the energy for adaptivity is increased. As shown, the increased energy demand is less than the decreased energy, which results an energy saving in total. However, the increase of energy of adaptivity is resulted by the possibility to deactivate software components and ECUs. As shown in Table VI most of the energy is necessary for (de)activation of ECUs, which enables the saving of the ECU offset energy demand. It is mentionable that just the system context "Vehicle Speed" is evaluated in this case study and just two applications are available. Within real vehicles a lot more applications are executed (cf. Section III) and more context information is interpretable. Through that a lot of system states are possible, which may result a lot of energy to enable adaptivity. Even within the 1180 seconds of the NDEC the system has to switch the states 48-times. Through that, it is necessary to model and evaluate the energy demands of different allocations to evaluate and reach a trade-off. However, a deactivation of components to save energy is only useful, if more energy is saved than needed for the deactivation and activation process [2].

TABLE V. ESTIMATED ENERGY DEMAND OF THE SOFTWARE COMPONENT ALLOCATIONS "FUNCTION-BASED ALLOCATION" AND "ENERGY-FOCUSED ALLOCATION"

	Function-based		Energy-focused			
	Energy Demand	Percentage	Energy Demand	Percentage	Difference	
System	20725.11 Ws		16989.03 Ws		-3736.08 Ws	-18.03 %
CPU	3441.16 Ws	16.61 %	3330.85 Ws	19.61 %	-110.31 Ws	-3.21 %
ECU Offset	9372.00 Ws	45.22 %	8229.00 Ws	48.44 %	-1143.00 Ws	-12.20 %
Sensors/Actuators	4711.41 Ws	22.73 %	3570.02 Ws	21.01 %	-1141.39 Ws	-24.23 %
Communication	2933.12 Ws	14.15 %	1526.24 Ws	8.98 %	-1406.88 Ws	-47.97 %
Adaptivity	267.42 Ws	1.29 %	332.92 Ws	1.96 %	+65.50 Ws	+24.49 %

TABLE VI. ENERGY DEMAND IN DETAIL FOR COMMUNICATION AND ADAPTIVITY OF THE "FUNCTION-BASED ALLOCATION" AND THE "ENERGY-FOCUSED ALLOCATION"

	Function-based		Energy-focused	
	Energy Demand	Percentage	Energy Demand	Percentage
Communication	2933.12 Ws		1526.24 Ws	
Comm. Connections (incl. Transfer)	2091.21 Ws	71.30 %	1052.30 Ws	68.95 %
Transfer	11.50 Ws	0.39 %	13.37 Ws	0.88 %
Listener	838.84 Ws	28.60 %	469.02 Ws	30.73 %
Energy Saving Mode	0.10 Ws	0.10 %	4.93 Ws	0.32 %
Adaptivity	267.42 Ws		332.93 Ws	
(De)Activation of Software Components	2.42 Ws	0.91 %	2.92 Ws	0.88 %
(De)Activation of ECUs	265.00 Ws	99.09 %	330.00 Ws	99.12 %

The resulted energy demand of the case study also shows that the energy savings are realized by the ECU, sensor/actuators and communication instead of the CPU. Today's CPUs are highly optimized and further energy saving potential is difficult to reach. In addition, the percentage of different energy demands are more or less equal between the two allocations. This means that it is necessary to consider the entire system, instead of focusing a specific part of the system. The case study shows the relevance of modeling the energy demand of adaptive, networked, embedded systems as a whole. The accuracy of the estimation depends on the considered system. Using the accuracy ranges presented within the previous work concerning the energy model (cf. Section II) the accuracy is theoretically between -11.4 % and +7.1 %. (The estimation accuracy of ECU offset and adaptivity energy demand is estimated to be ± 10 %.)

In the following section, the results are concluded.

VII. Conclusion

This paper presented a case study, which shows the applicability of the energy model, which was presented within previous work, and demonstrates the possible energy saving potential within adaptive, networked, embedded systems. The energy model is part of previous work and was used within this case study to estimate the energy demand of two allocations of software components.

Within the case study of this paper two advanced driver assistant systems were presented in detail including the resulting software architecture and the necessary software components. The considered system is adaptive and deactivates temporarily unnecessary software and hardware components which are not needed during the current system context. Within this case study the vehicle speed was used as the system context, which was derived by the New European Driving Cycle. Different speed thresholds result the activation or deactivation of (un)necessary components. The considered hardware architecture and the energy relevant parameters were presented. Afterwards, a simulation was briefly explained, which enables the estimation of the energy demand of an adaptive, networked, embedded system during a specific system context. Finally, the estimated energy demand of the two software component allocations were presented and the results discussed. The simulation results showed an energy saving potential of about 18 %. The energy demand of the adaptivity process itself increases significantly, nevertheless this is still less than the enabled energy saving of the other components resulted by adaptivity.

The results of this paper showed the relevance of modeling the energy demand of adaptive, networked, embedded systems and a considerable energy saving potential. This kind of estimation at the network level needs an appropriate energy model to estimate the energy demand of adaptive, networked, embedded systems during the development of the system, which was presented within previous work.

Acknowledgment

The research leading to these results has received funding from the German Federal Ministry for Economic Affairs and Energy (BMWi).

References

[1] A. Barthels, J. Fröschl, H.-U. Michel, and U. Baumgarten, "An Architecture for Power Management in Automotive Systems," in Proceedings of the International Conference on Architecture of Computing Systems, vol. 7179, Springer, 2012, pp. 63–73, ISBN: 978-3-642-28292-8.

[2] C. Schmutzler, A. Krüger, F. Schuster, and M. Simons, "Energy efficient automotive networks: state of the art and challenges ahead," in International Journal of Communication Networks and Distributed Systems, vol. 9, no. 3/4, 2012, pp. 266–285, ISSN: 1754-3916.

[3] J. Weber, "Automotive Development Processes: Processes for Successful Customer Oriented Vehicle Development," Springer-Verlag Berlin Heidelberg, 2009, ISBN: 978-3-642-01253-2.

[4] A. Monetti, T. Otter, and N. Ulshöfer, "Spritverbrauch senken, Reichweite erhöhen: System-Basis-Chip für den Teilnetzbetrieb am CAN-Bus," Elektronik Automotive, no. 11, 2011, pp. 24–27, ISSN: 1614-0125.

[5] A. D. Little, "Market and Technology Study Automotive Power Electronics 2015," 2006, URL: www.adlittle.com/downloads/tx_adlreports/ADL_Study_Power_Electronics_2015.pdf [accessed: 2016-05-17].

[6] P. Heinrich and C. Prehofer, "Network-Wide Energy Optimization for Adaptive Embedded Systems," ACM SIGBED Review, vol. 10, no. 1, 2013, pp. 33–36, ISSN: 1551-3688.

[7] P. Heinrich and C. Prehofer, "Early Energy Estimation in the Design Process of Networked Embedded Systems," in Proceedings of the 3rd International Conference on Pervasive Embedded Computing and Communication Systems, 2013, pp. 214–220, ISBN: 978-989-8565-43-3.

[8] P. Heinrich, H. Bergler, and D. Eilers, "Energy Consumption Estimation of Software Components based on Program Flowcharts," in Proceedings of the 11th IEEE International Conference on Embedded Software and Systems (ICESS), 2014, pp. 550-553, ISBN: 978-1-4799-6122-1.

[9] P. Heinrich, H. Bergler, and E. Oswald, "Early Energy Estimation of Networked Embedded Systems Executing Concurrent Software Components," in International Journal of Modeling and Optimization, vol. 5, no. 2, pp. 119–127, 2015, ISSN: 2010-3697.

[10] P. Heinrich, D. Gossen, E. Oswald, and R. Knorr, "Early Energy Estimation of Heterogeneous Embedded Networks within Adaptive Systems," in Energy Efficient Vehicles 2015, B. Bäker and L. Morawietz, Eds., Dresden: TUDpress, 2015, pp. 64–76, ISBN: 978-395-9080-08-8.

[11] P. Marwedel, "Embedded System Design: Embedded Systems Foundations of Cyber-Physical Systems," 2nd ed., Springer, 2011, ISBN: 978-940-0702-56-1.

[12] A. Metzner, "Effizienter Entwurf verteilter eingebetteter Echtzeit-Systeme," Dissertation, Universität Oldenburg, 2007.

[13] AUDI AG, "Adaptive Cruise Control with Stop & Go Function," 2011, URL: www.audi-technology-portal.de/en/electrics-electronics/driver-assistant-systems/adaptive-cruise-control-with-stop-go-function [accessed: 2016-05-17].

[14] Robert Bosch GmbH, "Kraftfahrtechnisches Taschenbuch," 26th ed., Wiesbaden: Vieweg, 2007, ISBN: 978-383-4814-40-1.

[15] BMW AG, Active Cruise Control with Stop&Go Function, 2015, URL: www.bmw.com/com/de/insights/technology/technology_guide/ articles/active_cruise_control_stop_go.html [accessed: 2016-05-17].

[16] U.S. Software System Safety Working Group, "Adaptive Cruise Control System Overview," 2005, URL: www.sunnyday.mit.edu/safety-club/workshop5/Adaptive_Cruise_Control_Sys_Overview.pdf [accessed: 2016-05-17].

[17] H.-H. Braess and U. Seiffert, Eds., "Vieweg-Handbuch Kraftfahrzeugtechnik," 5th ed., ser. ATZ-MTZ-Fachbuch, Wiesbaden: Vieweg, 2007, ISBN: 978-383-4802-22-4.

[18] Daimler AG, Active Parking Assist, 2016, URL: www.techcenter.mercedes-benz.com/en/active_parking/detail.html [accessed: 2016-05-17].

[19] H. Hansson, M. AAkerholm, I. Crnkovic, and M. Torngren, "SaveCCM - a component model for safety-critical real-time systems," in Proceedings of the 30th EUROMICRO Conference, 2004, pp. 627–635, ISBN: 978-076-9521-99-2.

[20] A. G. Fontquerni, "Embedded Linux RADAR device: Taking advantage on Linaro tools and HTML5 AJAX real-time visualization," Embedded Linux Conference Europe, Barcelona, 2012, URL www.elinux.org/images/7/75/Embedded_Linux_RADAR_Device.pdf [accessed: 2016-05-17].

[21] Freescale Semiconductor, "MPC5674F Microcontroller Data Sheet," 2015, URL: www.cache.freescale.com/files/32bit/doc/data_sheet/MPC5674F.pdf [accessed: 2016-05-17].

[22] T. Weber, V. Lauer, D. Mann, and M. Simons, "Das umfassende Energiemanagement: Vom konventionelle Verbrenner bis zum E-Antrieb," in 4. VDI-Tagung Baden-Baden Spezial 2010, ser. VDI-Berichte, vol. 2098, VDI Verlag, 2010, ISBN: 978-3-18-0920-98-6.

[23] A. Y. Benbasat, "An Automated Framework for Power-Efficient Detection in Embedded Sensor Systems," Dissertation, 2007, URI: http://hdl.handle.net/1721.1/38524.

[24] Robert Bosch GmbH, "Fernbereichsradarsensor LRR3: Long-Range Radar, 3. Generation," 2009, URL: www.produkte.bosch-mobility-solutions.de/media/db_application/pdf_2/de/LRR3_Datenblatt_DE_2009.pdf [accessed: 2016-05-17].

[25] Continental AG, "Short Range Radar Sensor SRR 20X /-2 /-2C /-21," 2013, URL: www.conti-online.com/www/industrial_sensors_de_en/themes/srr20x_en.html [accessed: 2016-05-17].

[26] Freescale Semiconductor, "Automotive Engine Control IC: Technical Data," 2014, URL: www.cache.freescale.com/files/analog/doc/data_sheet/MC33810.pdf?pspll=1 [accessed: 2016-05-17].

[27] NXP Semiconductors, "KMI17/4 - Rotational speed sensor: Product data sheet: Rev. 1, September 2014," 2014, URL: www.nxp.com/documents/data_sheet/KMI17_4.pdf [accessed: 2016-05-17].

[28] NXP Semiconductors, "KMA220 - Dual channel programmable angle sensor: Product data sheet: Rev. 2, April 2013," 2013, URl: www.nxp.com/documents/data_sheet/KMA220.pdf [accessed: 2016-05-17].

[29] Robert Bosch GmbH, "Combined inertial sensor for vehicle dynamics control - SMI650: Automotive Electronics," 2013, URL: www.bosch- semiconductors.de/media/pdf_1/einzeldownloads/vehicle dynamics_systems/datenblatt_smi650.pdf [accessed: 2016-05-17].

[30] Robert Bosch GmbH, "CAN Specification: Version 2.0," Stuttgart, 1991, URL: www.bosch-semiconductors.de/media/ubk_semiconductors/pdf_1/canliteratur/can2spec.pdf [accessed: 2016-05-17].

[31] Microchip Technology Inc., "High-Speed CAN Transceiver: MCP2561/2: Revision C," 2014, URL: www.microchip.com/downloads/en/DeviceDoc/20005167C.pdf [accessed: 2016-05-17].

[32] Microchip Technology Inc., "Stand-Alone CAN Controller with SPI Interface: MCP2515: Revision G," 2012, URL: www.microchip.com/downloads/en/DeviceDoc/21801G.pdf [accessed: 2016-05-17].

[33] Standard Microsystems Corp., "Small Footprint MII/RMII 10/100 Ethernet Transceiver with HP Auto-MDIX and flexPWR Technology: LAN8710A/LAN8710Ai: Revision 1.4," 2012, URL: www.microchip.com/downloads/en/DeviceDoc/8710a.pdf [accessed: 2016-05-17].

[34] Cirrus Logic Inc., "CS8900A: Product Data Sheet: Crystal LAN Ethernet Controller," 2010, URL: www.cirrus.com/jp/pubs/proDatasheet/CS8900A_F5.pdf [accessed: 2016-05-17].

[35] C. Schmutzler, "Hardwaregestützte Energieoptimierung von Elektrik/Elektronik-Architekturen durch adaptive Abschaltung von verteilten, eingebetteten Systemen," Dissertation, 2012, ISBN: 978-86-6448-75-9.

[36] European Parliament, Council of the European Union, "Directive 98/69/EC of the European Parliament and of the Council," 13.10.1998, URL: www.eur-lex.europa.eu/LexUriServ/LexUriServ.do?uri=CONSLEG:1998L0069:19981228:EN:PDF [accessed: 2016-05-17].

[37] S. Schmidt, "EcoTest: Testing and Assessment Protocol: Release 2.0," 2015, URL: www.ecotest.eu/html/testing%20and%20assessment%20protocol.pdf [accessed: 2016-05-17].

[38] J. Banks, J. S. Carson, II, B. L. Nelson, and D. M. Nicol, "Discrete-Event System Simulation," 4th ed., ser. Prentice-Hall international series in industrial and systems engineering, Pearson Prentice Hall, 2005, ISBN: 978-013-6062-12-7.

[39] P. Bastian, "Grundlagen der Modellbildung und Simulation," Vorlesungsskript, Universität Stuttgart, Stuttgart, 2008, URL: www.conan.iwr.uni-heidelberg.de/teaching/scripts/msarticle.pdf [accessed: 2016-05-17].

[40] H.-J. Bungartz, S. Zimmer, M. Buchholz, and D. Pflüger, "Modeling and Simulation: An Application-Oriented Introduction," Berlin: Springer, 2013, ISBN 978-364-2395-24-6.

[41] B. P. Zeigler, H. Praehofer, and T. G. Kim, "Theory of Modeling and Simulation: Integrating Discrete Event and Continuous Complex Dynamic Systems," 2nd ed., Academic Press, 2000, ISBN: 978-012-7784-55-7.

[42] G. Walla, A. Herkersdorf, A. S. Enger, A. Barthels, and H.-U. Michel, "An automotive specific MILP model targeting power-aware function partitioning," in Proceedings of the 2014 International Conference on Embedded Computer Systems: Architectures, Modeling, and Simulation (SAMOS XIV), pp. 299–306, 2014, ISBN: 978-147-9937-69-1.

[43] R. Jejurikar and R. Gupta, "Dynamic voltage scaling for systemwide energy minimization in real-time embedded systems," in Proceedings of the 2004 International Symposium on Low Power Electronics and Design, New York, pp. 78–81, 2004, ISBN: 978-158-1139-29-7.

[44] AUTOSAR Development Cooperation, "Automotive Open System Architecture," URL: www.autosar.org [accessed: 2016-05-17].

The TUCool Project - Low-cost, Energy-efficient Cooling for Conventional Data Centres

Matthias Vodel* and Marc Ritter*
*Chemnitz University of Technology
*Chemnitz, Germany

email: [1]matthias.vodel@hrz.tu-chemmnitz.de,
email: [2]marc.ritter@cs.tu-chemmnitz.de

***Abstract*—Air-conditioned cooling concepts still represent the usual cooling solution for thousands of mid-sized data centres. These data centres consists of different types of IT-infrastructure components, as well as different hardware generations. During the last decade, the optimisation regarding energy-efficiency for such central IT locations becomes one of the most important challenges. Green-IT improvements for existing data centres means an adaptive and safe parametrisation of the air-conditioning-system and its control mechanisms. But in order to handle these issues, the available amount of sensor data is critical. A large diversity of distributed sensor devices allows a more precise system management. In this context, the TU Chemnitz develops a cost-efficient and smart solution to improve the sensor knowledge base as well as the control mechanisms. We are using local sensor capabilities within the hardware components and combine these information with actual system loads to create an extended knowledge base, which also provides adaptive learning features. In a first research stage, we analyse the actual cooling environment and measure several operational scenarios for creating a detailed simulation model. The respective results demonstrates a huge optimisation potential. Accordingly, an optimised trade-off between power consumption and cooling capacity may result in significant cost savings.**

Keywords - air-conditioning; data centre; optimisation; energy-efficiency; adaptive; sensor fusion; control loop; control system.

I. Introduction

Traditional data centres are characterised by heterogeneous hardware components and generations. Multiple hardware generations over several decades are running side-by-side. Such locations include all kinds of IT-infrastructure like storage systems, network core components, and server systems. Due to this mixed environment, compromises regarding the cooling capabilities are necessary. Due to physical limitations regarding cooling power and energy density per rack, a large amount of space capacity inside each air-cooled server rack is wasted [1][2]. Accordingly, the optimisation of such traditional air-cooled data centre environments regarding energy- and cost-efficiency is one of the central challenges for hundreds of institutions in the public and educational domain [3].

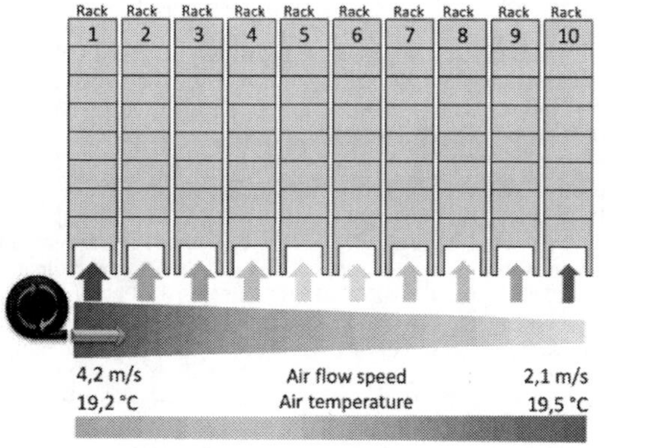

Figure 1. Key problems for traditional, air-cooled data centre environments. In-homogeneous air temperature and air flow speed dependent on the positioning of the server rack. Starting from the air intake on the left side, the cooling capacity shrinks from rack to rack [4].

II. Problem Description

There are two major problems for usual air-cooled data centres: *Inhomogeneous air temperature* and the *inhomogeneous air flow* inside the data centre. These parameters are strongly dependent on the server rack location within the room and even on the position of each individual server component inside the rack. These two challenges are shown in Figure 1 based on measurements in our TU Chemnitz data centre.

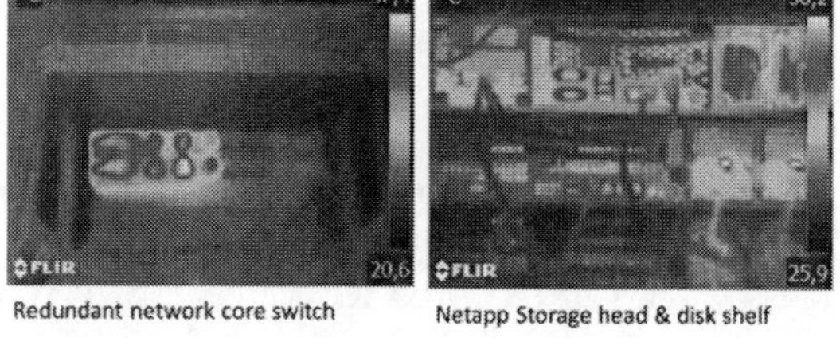

Figure 2. Detailed heat analysis for the air offtake behaviour in different hardware components.

With focus on an entire data centre with multiple server racks and hundreds of server systems, an additional issue becomes critical: Turbulences and interferences between different air flows around the individual racks. These effects have a huge impact on the cooling efficiency. In order to quantify this impact, we are analysing detailed heat images for each hardware component. The heat distribution and model-specific air flow characteristic results in individual heat patterns for each component class. Figure 2 visualises such a component-based heat analysis for two different air offtakes.

Facing these efficiency challenges from an administrative perspective, the monitoring and measuring of the respective values appears in a very basic manner [5][6]. Usual data centre environments only provide a few global temperature sensors for the entire room. Accordingly, the control loop for the air conditioning is very simple. Besides the global room temperature, no further information are available.

III. Related Work

Due to these issues, several professional solutions try optimise this situation regarding monitoring capabilities, sensor data sources, management & control processes as well as cost- and energy savings.

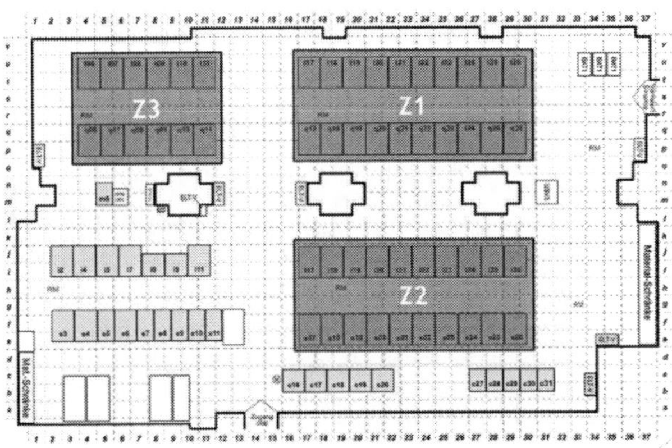

Figure 3. TU Chemnitz data centre with three cold aisle containments, which represent the operational zones Z1, Z2, and Z3.

A. *Cold Aisle Containment & Air Boosters*

One of the most efficient optimisation steps for traditional air-cooled data centres represents *cold aisle containments*, which allows us to concentrate the cooling capacity directly to the server hot spots within the room. Accordingly, we reduce the effective volume from the entire room space to single enclosures with a significant smaller capacity. Figure 3 shows the three realised cold aisle containments of the TU Chemnitz data centre.

Each containment provides individual temperature sensors and is equipped with optional *booster* elements. The booster technology is shown in Figure 4. As one can see, the boosters allow us to modify the air flow individually for each zone. In order to establish such cold aisle containments, each hardware component has to be re-organised regarding the direction of the air flow. Air intakes have to be located inside the containment, air offtakes outside the enclosures. Accordingly, the installation of these containments is very time-consuming, requires a detailed timeline and is critical with respect to system downtimes or failures.

But anyway, the control cycles as well as the information database for adapting the boosters and the air conditioning system are still the same. The control loops only operate in a static, reactive approach, based on single temperature measurements inside the containment. No further information are available.

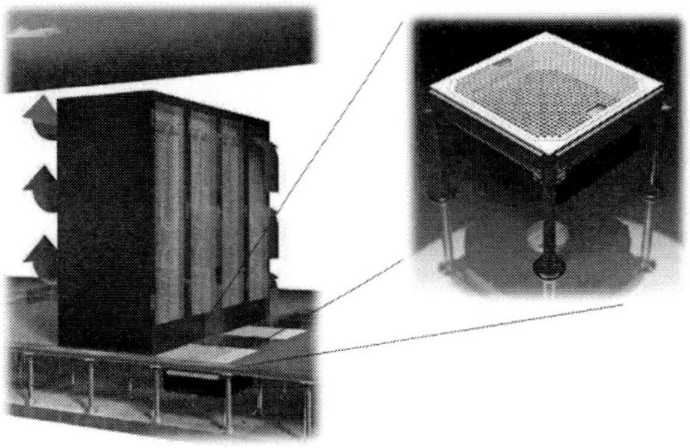

Figure 4. Booster components for dynamic adaptation of the air flow in different, individual housing areas [4].

B. *Genome Project*

In order to provide a better sensor data knowledge base, Microsoft Research starts the *Genome project* [7][8], which adds dedicated wireless sensor nodes to each server rack. These nodes (called *Genomotes*, see Figure 5) are organised in a master-slave chained sensor network design (*RACNet*), based on the IEEE 802.15.4 low-power, low-data rate communication stack [9]. The RACNet infrastructure provides several information sets about the environmental status, including heat distribution, hot spots, and facility layout. Each node sends its data to a predefined data sink, which creates a global view regarding the health status. The entire raw data is merged together for different data representation tasks (analysis, prediction, optimisation, and scheduling).

C. *SynapSense*

Another company, which also uses such kinds of sensor nodes is *SynapSense* [10]. Here, several node classes with different types of information are available, e.g., *Therma Nodes* (see Figure 5), *Pressure Nodes*, or *Constellation Nodes*. The data sets from the nodes are processed in a centralised manner by a special software tool, which is able to adapt and to steer the air conditioning system.

All of these approaches possess two critical disadvantages. The first one deals with additional hardware costs for the different sensors. This includes costs for installation, configuration, operation, and maintenance. For large-scaled data centre environments, the required financial resources are very high [11]. The second disadvantage represents the type of data. All of these systems are measuring external parameters from the current point in time, thus providing no learning capability from the past. In addition, there are no server-internal data sources like the system load or any kind of hardware health status as well.

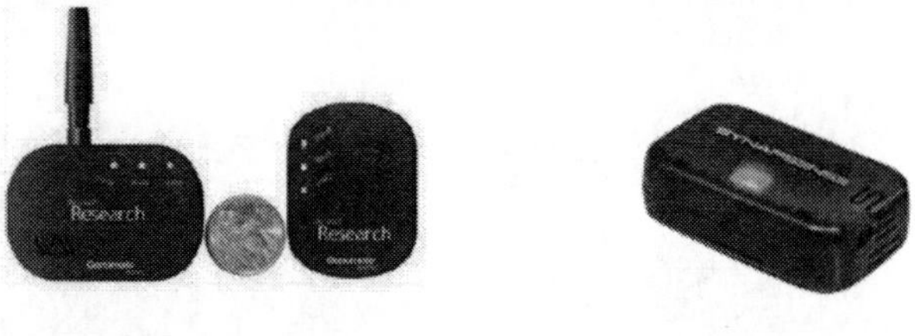

Figure 5. Sensor nodes for data centre monitoring. Left: *Genomotes* from Microsoft Research; Right: *ThermaNode* from SynapSense

IV. Sensor Data & System Management

Our idea for a more efficient solution is very simple but quite efficient. With our monitoring and control approach, we include different software-based sensor plugins. Each plugin represents a class module for a specific kind of sensor data. In contrast to other approaches, we are using local sensor capabilities of each hardware components inside the data centre. E.g., a given server system typically provides several temperature sensors, located at the mainboard, the CPUs, and the cooling fans (illustrated in Figure 6).

Further information modules are monitoring and learning the system load values of physical/virtual server entities and the respective impact on the data centre temperature behaviour. We are mapping all of these temperature and system load information into one extended knowledge base. Furthermore, different sensor data sources are merged together to more abstract information sets. The fusion results indicate the actual health status of the data centre as well as a prediction trend for the future. Past monitoring data represents a continuous input for machine learning capabilities. In order to save energy and costs, a feasible prediction model [12] is necessary for adapting the cooling power.

Temperature peaks for short term loads and local hot spots are handled with an increased air flow, which means local air booster elements. Such short term situations include hundreds of boot processes of virtual desktops in the morning or backup tasks in the night. Also, small- and mid-size compute jobs for cluster installations may result in such short term temperature peaks.

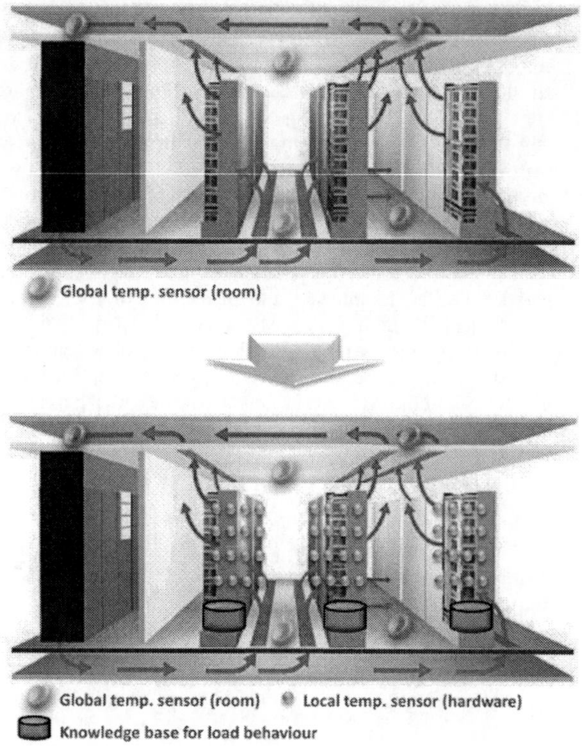

Figure 6. Extension of the knowledge base by using additional sensors and load data from the individual server systems [4].

On the other side, the control system must handle the long term temperature behaviour inside the data centre, e.g., the differences between working days and weekends as well as day & night periods. For such scenarios, the entire air conditioning system with its specific cooling capacity has to be adapted periodically.

V. Analysis & Results

In this paper, we analysed three different impact parameters, based on our data centre environment with three cold and one warm zone.

First of all, we measured the temperature differences between each zone and we identified noticeable influences between the encapsulated cold aisles. Dependent on the positioning of each server system within the zones, nearby area are heated up significantly (see Figure 7). A very simple but efficient solution represents any kind of obstacle between the cold zones. Accordingly, the air interferences are minimised and the temperature impact decreases.

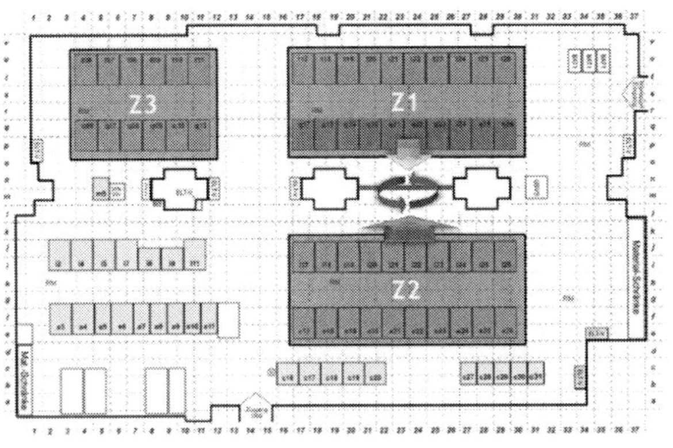

Figure 7. Impact of nearby cold aisles dependent on the hardware location. Indirect temperature increase for isolated cold aisle containments. The red line represents any kind of separating obstacle between these zones.

A second result deals with the hardware diversity and the hardware distribution within the data centre. We found out that a mixed positioning of different hardware types (storage, compute, network) in the available zones is much more energy-efficient than an organised positioning. If we put all the compute and server systems in one zone, we create hot spots regarding the temperature. Due to the fact that we only have one global air conditioning system, we have to increase the basic cooling capacity (and the respective power consumption) for cooling down only one critical zone.

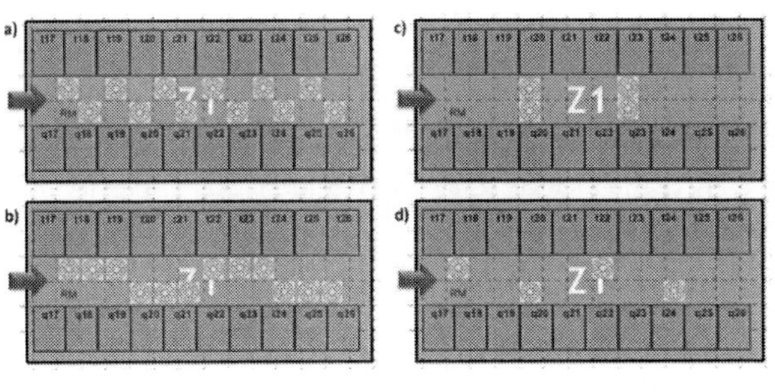

Figure 8. Different booster configurations in a given cold aisle containment.

A third optimisation result focuses on the booster technology. In order to find an optimal trade-off between hardware efforts and cooling efficiency, the amount of booster elements is critical. But there is no direct linear relation between the amount of boosters and the cooling capacity. Furthermore, the booster locations are relevant. Figure 8 visualises this topic.

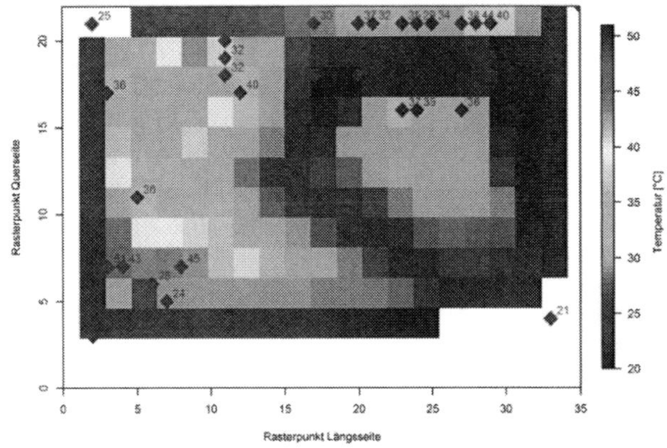

Figure 9. TU Chemnitz data center heat map [4]. Hot spots without cold aisle containment in the bottom left corner are clearly visible.

Type a) and b) is very hardware-extensive but the benefits in comparison to type d) are minimal. In contrast, the cooling efficiency of type d) is significant better than type c) while using the same amount of boosters. Accordingly, the well-organised booster positioning allows excellent cooling capabilities without massive hardware efforts.

All measurements in our data centre are monitored over a time period of three months. We visualise the results in complex heat maps as shown in Figure 9. The sample rate for all temperature measurements is 1 data set per minute, leading to more than 100000 data points per sensor. Server-internal system load measurements are monitored with 1 data set per second to calculate a feasible average value per minute. Accordingly, we map global and local temperature values together with local system loads and fan speed information. Hence, the merged data is available in one central knowledge base to allow a detailed and situation-specific adaptation of the cooling capacity.

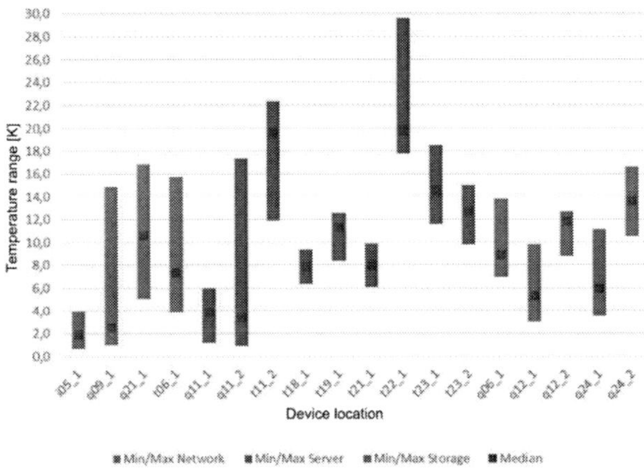

Figure 10. Measured temperature ranges for several hardware components at different locations (coordinates referenced in Figure 3).

Finally, we illustrate the operational temperature ranges for different hardware components during the tests. Figure 10 shows minimum, maximum, and average values.

VI. Conclusion

In this paper, we analysed the energy-efficiency of heterogeneous data centre environments with traditional air cooling systems. We introduced a cost-efficient and smart approach for improving the sensor data in order to implement an adaptive management system for the cooling capacity. The proposed concept does not require any further hardware components or installation efforts. The system utilizes given sensor sources from each hardware system and aggregates these data sets into one single knowledge base. Furthermore, we also discuss the impact of weak parameters like the hardware positioning inside the data centre and its cold aisles. Also, the booster configuration, as well as influences between different cold and warm zones were analysed. As already mentioned, improved control cycles for the cooling systems allow us to reduce the basic cooling level. Short term temperature peaks can be avoided with a local adaptation of the air flow using booster components. Accordingly, we are able to save massive energy and costs without any increased risk level for the hardware [4]. In this context, the presented results offer high potential for optimisations. The research goal deals with the vision of optimising an entire data centre environment based on extensive simulation processes without any trial-and-error approaches on the real hardware.

Acknowledgement

RenewIT

This research work is supported by the *RenewIT* project, which is co-financed by the 7th Framework programme of the European community (FP7) under grant agreement number 608679. All measurements were done within the data centre environment of the *Technische Universität Chemnitz*.

References

[1] R. Lent. Evaluating the cooling and computing energy demand of a datacentre with optimal server provisioning. *Future Generation Computer Systems 57*, pages 1–12, April 2016.

[2] Y. Wang G. Xing X. Wang X. Wang B. Punch J. Chen, R. Tan and D. Colbry. Sensor system for high-fidelity temperature distribution forecasting in data centers. *ACM Trans. Sen. Netw. 11, 2, Article 30*, page 25 ff., December 2014.

[3] R. Guerra G. Zanatta, G. D. Bottari and J. Leite. Building a wsn infrastructure with cots components for the thermal monitoring of datacenters. In *In Proceedings of the 29th Annual ACM Symposium on Applied Computing (SAC '14)*, pages 1443–1448, 2014.

[4] M. Vodel and M. Ritter. Adaptive Sensor Data Fusion for Efficient Climate Control Systems. In *Proc. of HCII 2015 - International Conference on Human-Computer Interaction*, pages 1–11. Springer, 2015.

[5] J. Liu and A. Terzis. Sensing data centers for energy efficiency. *Philosophical Transactions of The Royal Society A*, pages 136–157, January 2012.

[6] M. Vodel. *Energy-efficient communication in distributed, embedded systems*. PhD thesis, TU Chemnitz, Chemnitz, February 2014. ISBN 978-3-941003-18-7.

[7] Microsoft Research. DC Genome Project, 2009. [Online; last access May 2016].

[8] US Dept. of Energy. Wireless sensors improve data center energy efficiency. Technical Report CSO 20029, USA, September 2010.

[9] L. Luo A. Terzis C.-J. Mike Liang, J. Liu and F. Zhao. RACNet: A High-Fidelity Data Center Sensing Network. In *Proc. of the SenSys (ACM Conference on Embedded Network Sensor Systems)*, pages 9–12. ACM, November 2009.

[10] SynapSense. Active Control. http://www.synapsense.com, 2015. [Online; last access May 2016].

[11] Y. Jia K. Yoshii R. Ross M. G. Rodriguez, L. E. Ortiz Uriarte and P.H. Beckman. Wireless sensor network for data-center environmental monitoring. In *Proc. of the 5th International Conference on Sensing Technology (ICST)*, pages 533–537, Nov 2011.

[12] J. Liu S. Nath A. Terzis L. Li, C-J. Mike Liang and C. Faloutsos. ThermoCast: A Cyber-Physical Forecasting Model for Data Centers. In *Proc. of the SIGKDD (Conference on Knowledge Discovery and Data Mining)*, pages 13–15. ACM, August 2011.

Ensuring Secure Communication in Critical Infrastructures

Steffen Fries and Rainer Falk
Corporate Technology
Siemens AG
Munich, Germany
e-mail: {steffen.fries|rainer.falk}@siemens.com

Abstract—**Critical infrastructures as backbone of the society and economy are increasingly the target of cyber attacks. These infrastructures have been isolated in the past, but are connected more and more also with external systems to allow for new and combined services. This immediately requires the protection of the communication connections to external sites. Legislation and operation have taken this into account and provide the necessary framework for posing specific communication security requirements. From the technical side, different security counter measures exist to cope with the given requirements, but it has to be ensured that these technical means are not only provided, but in fact applied in operation. This paper describes a new approach to ensure that during the setup of a secure communication connection the appropriate security is effectively negotiated with respect to permissible cipher suites for authentication, message integrity, and confidentiality. The application within a Smart Grid is used as example application domain.**

Keywords–security; critical infrastructure; smart energy grid; industrial automation; Internet of Things; secure communication; security policy; security protocol; Transport Layer Security

I. Introduction

Critical Infrastructures (CI) and especially cyber security in critical infrastructures has gained more momentum over the last years. The term "critical infrastructure" in the context of this paper is used to describe technical installations, which are essential for the functioning of the society and economy of a country, but also globally. Typical critical infrastructures in this context are the smart energy grid (including central or distributed energy generation, transmission, and distribution), water supply, healthcare, transportation, telecommunication services, just to state a few. The increased threat level becomes visible, e.g., through reported attacks on critical infrastructure, but also through legislation, which meanwhile explicitly requires the protection of critical infrastructures and reporting about serious attacks.

Information Technology (IT) security in the past was addressed mostly in common enterprise IT environments, but there is a clear trend to provide more connectivity to operational sites, which are quite often part of the critical infrastructure. Examples for operational sites are industrial automation or energy automation. This increased connectivity leads to a tighter integration of IT and Operational Technology (OT). IT security in this context evolves to cyber security to underline the mutual relation between the security and physical effects.

This paper focuses on the smart energy grid as example for critical infrastructures. The smart energy grid consists of several interworking parts depending on communication in a secure and reliable way. These parts are given through the classical power system elements like a centralized power generation, power transmission (typically high voltage and wide area connections), power distribution (low and medium voltage) and the consumer at the end of the supply chain. In the last years, the usage of renewable energy, e.g., through solar cells or wind power, became increasingly important to generate environmentally sustainable energy and thus to reduce greenhouse gases leading to global warming. Utilizing renewable energy in the power grid can be achieved in basically two ways: replacing classical power plants with renewable power plants likewise connected to the transmission grid. Alternatively, Decentralized Energy Resources (DER) are connected to the distribution network. In both cases, the energy generation through a grid of renewables needs to be monitored and controlled to a similar level as in today's centralized energy generation by power plants, while utilizing widely distributed communication networks. DER may also be aggregated virtually on a higher level to build a virtual power plant (VPP). A VPP may be viewed from the outside in a similar way as a common power plant with respect to energy generation. But due to its decentralized nature, the demands on communication necessary to control the VPP are much more challenging.

The target architecture for this paper is depicted on abstract level in Figure 1 below. It investigates into cyber security requirements from different sources providing specifics for secure communication and utilized technical security measures. Specifically, it proposes technical means to ensure the desired strength of security (given through a security policy) for the communication in the operation environment. The remainder of the paper is structured as follows. Section 2 investigates in cyber security requirements given through regulation, standards and guidelines. Section 3 investigates into Transport Layer Security (TLS) [1] as one common security protocol utilized power systems. Section 4 concentrates on the assurance that this security protocol is used with settings according to a given security policy. The technical proposal to achieve compliance to a given security policy for the communication between different entities of critical infrastructures is the main contribution of this paper. Note that this concept has not been implemented, yet. The conclusion discusses applicability to further security protocols and the necessity for an evaluation to determine the impact of the proposed solution to the overall system.

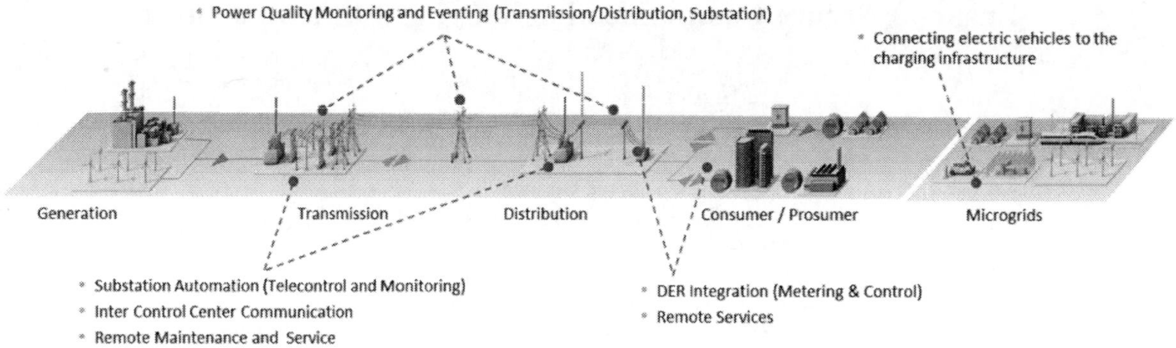

Figure 1. Overview Smart Energy Grid as Example for Critical Infrastructures

II. SMART ENERGY GRID SECURITY REQUIREMENTS

As stated in the introduction, the operational environment of critical infrastructures, in this paper on the example of the smart energy grid, differs from office environments or telecommunication environments in significant aspects. This leads to a different focus of general security requirements, like shown in the following Figure 2.

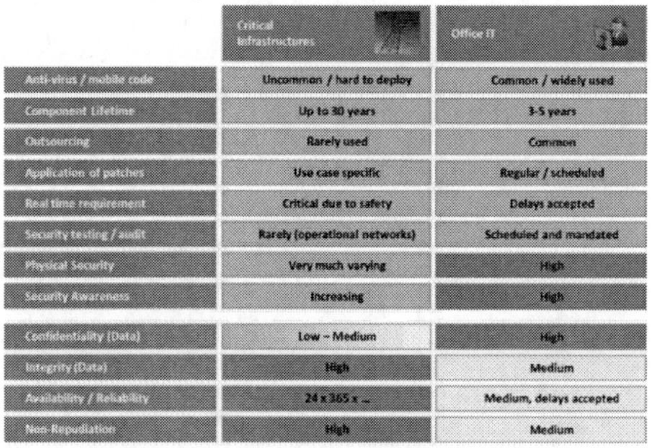

	Critical Infrastructures	Office IT
Anti-virus / mobile code	Uncommon / hard to deploy	Common / widely used
Component Lifetime	Up to 30 years	3-5 years
Outsourcing	Rarely used	Common
Application of patches	Use case specific	Regular / scheduled
Real time requirement	Critical due to safety	Delays accepted
Security testing / audit	Rarely (operational networks)	Scheduled and mandated
Physical Security	Very much varying	High
Security Awareness	Increasing	High
Confidentiality (Data)	Low – Medium	High
Integrity (Data)	High	Medium
Availability / Reliability	24 x 365 x ...	Medium, delays accepted
Non-Repudiation	High	Medium

Figure 2. Comparison CI and Office environment

These general security requirements are addressed in regulation, standards, guidelines and further customer specific or operator requirements. Figure 3 depicts example sources for such security requirements.

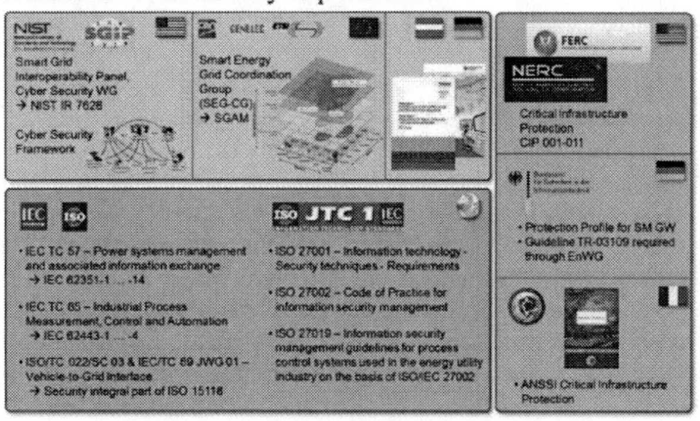

Figure 3. Sources for Security Requirements

As this paper focuses on communication security, the following subsections investigate into specific secure communication requirements in example requirement documents of different sources.

A. Regulative requirements

The regulative environments taken here as example focus on the operation of CI:

- The North American Electric Reliability Council (NERC) has established the Critical Infrastructure Protection (CIP) Cyber Security Standards CIP–002 through CIP–011 [2], which are designed as foundation of sound security practices across bulk power systems. They provide a consistent framework for security control perimeters and access management with incident reporting and recovery for critical cyber assets and cover functional, as well as non-functional requirements. NERC-CIP version 3 is formally controlled and enforced in the U.S. and in Canada. The first version originated in 2006 and has been continuously enhanced.

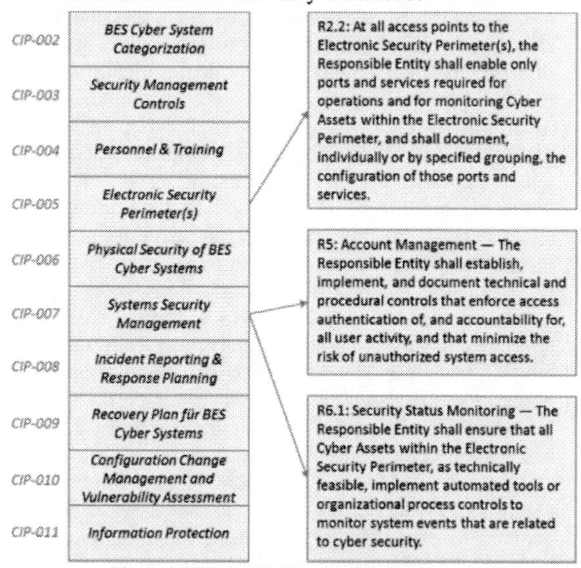

Figure 4. NERC-CIP Security Requirements

- A further example can be given by the legislation in Germany. Here, the IT security law has been finalized in 2015 requiring appropriate protection and monitoring, as well as reporting about security breaches for the operator of CI [3]. A specific regulation is the German Energy Act [4], which regulates in §21 the application of smart meters in facilities depending on the energy consumption/generation rate. The German "Bundesamt für Sicherheit in der Informationstechnik" (BSI) provides the technical guideline TR 03109 [5] to fulfill the requirements from the Energy Act and explicitly, how to ensure secure communication utilizing TLS to protect the communication.
- In France, the "Agence nationale de la sécurité des systèmes d'information" (ANSSI) regulates cyber security. Specifically, for secure communication there exists a guideline for the selection of TLS mechanisms providing appropriate protection [6].

The common approach of these regulations is, that they cover organizational requirements, process requirements and also technical requirements. The examples show that the security of communication is a clear part of the requirements.

B. Standards

Besides legislation, there exists a variety of standards, formulating security requirements or provide specific solutions to secure communication in an interoperable way. The following bullet list builds on the standards stated in Figure 3.

- IEC 62443, especially IEC 62443-3-3 [7]

  IEC 62443 is a security requirements framework defined in the IEC (International Electrotechnical Commission) Council) and can be applied to different automation domains, including energy automation, process automation, building automation, and others. As shown in Figure 5 the different parts are grouped into four clusters covering

  - common definitions, and metrics
  - requirements on setup of a security organization (ISMS related), and solution supplier and service provider processes
  - technical requirements and methodology on a secure system at system-wide level and
  - requirements to the secure development lifecycle of system components, and security requirements to such components at a technical level

  According to the methodology described in IEC 62443-3-2, a complex automation system is structured into zones that are connected by and communicate through so-called "conduits" that map for example to the logical network protocol communication between two zones. Moreover, this document defines Security Levels (SL) that correlate with the strength of a potential adversary. To reach a dedicated SL, dedicated requirements have to be met.

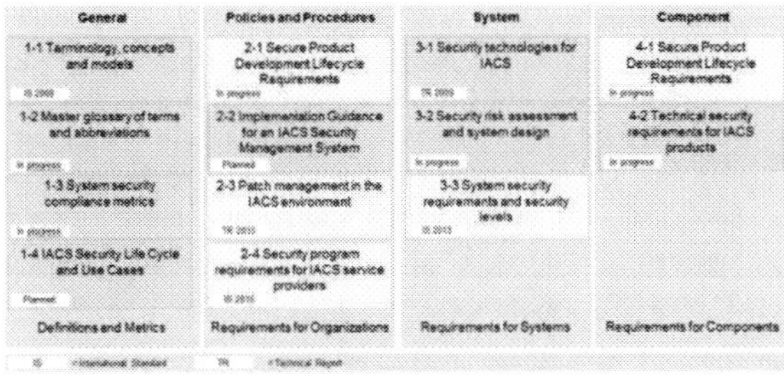

Figure 5. IEC 62443 Overview

Several requirements formulated in IEC 62443-3-3 [7] directly target communication security like:

  - Requirement 3.3.1 Communication integrity: "The control system shall provide the capability to protect the integrity of transmitted information".
  - Requirement 4.4.1 Communication confidentiality: "The control system shall provide the capability to protect the confidentiality of information at rest and remote access sessions traversing an untrusted network."

  These requirements are used her as an example that IEC 62443 requires the support of certain functionality. These requirements are linked to different security levels and thus have to be seen in the overall system context.

- IEC 62351, especially IEC 62351-3 [8]

  IEC 62351, which is also defined in the IEC, targets security mechanisms applicable to the power systems domain. The standard is split into different parts addressing specific security topics, as shown in Figure 6.

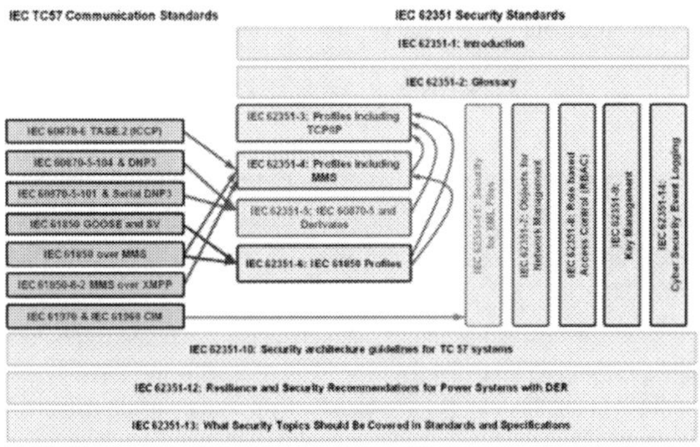

Figure 6. IEC 62351 Overview [8]

Specifically, IEC 62351-3 targets to secure TCP based communication by profiling the use of TLS and is referenced from other IEC 62351 parts. Profiling of TLS relates to narrowing available options in TLS like the requirement to utilize mutual authentication reducing the number of allowed algorithms or the disallowance of utilizing certain cipher suites, not providing sufficient protection. Moreover, this part also provides guidelines for utilizing options, which depend on the embedding environment. An example is the relation of using session renegotiation and session resumption in conjunction with the update interval of the certificate revocation information.

C. Guidelines

Besides regulations and standards, there also exist guidelines on how to address secure communication in specific application environments.

- The "Bundesverband für Energie- und Wasserwirtschaft" (BDEW) introduced a white paper defining basic security measures and requirements for IT-based control, automation and telecommunication systems for energy and water systems, taking into account general technical and operational conditions [10]. It can be seen as a further national approach targeting similar goals as NERC-CIP, but at a less detailed level. The white paper addresses requirements for vendors and manufacturers of power system management systems by directly relating to ISO 27000. Section 2.3 of this white paper focuses on communication and formulates specific requirements for integrity and confidentiality of connections.
- NISTIR 7628 [11] originates from the Smart Grid Interoperability Panel (Cyber Security WG) of the National Institute for Standards and Technology (NIST). It targets the development of a comprehensive set of cyber security requirements. The document consists of three subdocuments targeting strategy, security architecture, and requirements, and supportive analyses and references. It specifically formulates requirements for smart grid information system and communication protection.

III. TLS TO SECURE TCP COMMUNICATION

As shown in the previous section, there are numerous examples of requirements to secure communication, which leads to the necessity to be able to verify that the appropriate communication security is applied in fact in operational use. This section investigates into technical means to ensure secure communication by taking TLS as example, as it is used widely also in power automation systems (see IEC 62351 in section II.B) , to protect the communication.

TLS in its current version 1.2 defines protection means for TCP-based communication and is defined in Internet standard RFC 5246 [1]. Note, that the standard has a long history and is constantly being evolved to cope with new advances in cryptography and communication security. It supports a variety of authentication options for the communicating peers and allows the negotiation of the protection of the preceding communication in terms of integrity and confidentiality and also key management related options like key updates, etc. The combination of cryptographic algorithms for authentication, integrity, and confidentiality protection is called cipher suite. TLS is build upon several sub protocols that encapsulate the protocol operation in the different phases as shown in Figure 7.

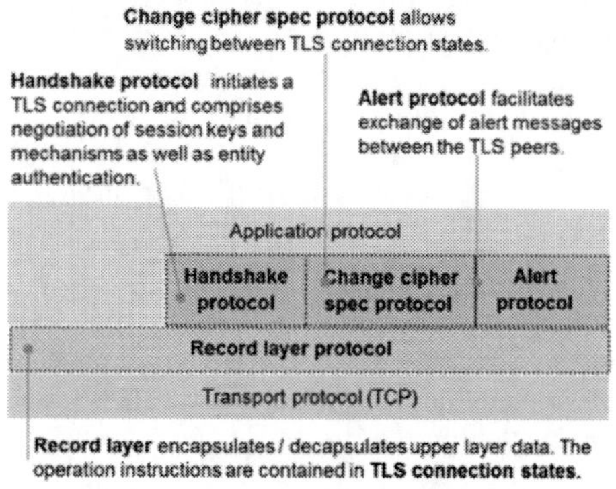

Figure 7. TLS Protocol Structure

For the discussion in this paper the most interesting phase is the TLS handshake, as it is performed in clear and allows the monitoring of the negotiated security options for the following communication session. Figure 8 shows the message exchange during the handshake.

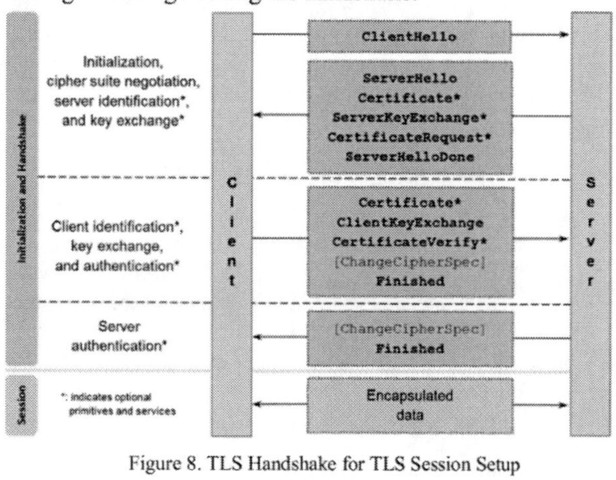

Figure 8. TLS Handshake for TLS Session Setup

Especially, the first phase of the handshake is in focus here, as it conveys the information for the cipher suite negotiation and the authentication of the communicating peers. In the *ClientHello* message, the client passes a list of cipher suites to the server containing the combinations of cryptographic algorithms supported in order of the client's preference. The server will then select a cipher suite and respond with a *ServerHello* message if a matching proposal was found. If no matching proposal was found, the server will issue a failure alert. Assumed that the server will authenticate towards the client, it will send its certificate as part other response. This allows the client to identify the server, validate the server certificate, as well as to utilize the server certificate during the further session key establishment. If the server additionally requires a client authentication as part of the TLS handshake, it will send a *CertificateRequest* message.

The second phase of the handshake targets the client identification (if requested) and the session key establishment and the authentication of both sides. The *Finished* message from the server to the client concludes the handshake and is the first message encrypted using the negotiated session key. It also contains a hash over the previously exchanged handshake messages to have a delayed verification of the integrity of the performed handshake.

Based on the provided TLS overview the handshake phase can be used to monitor the establishment of a secure communication, which can be audited by an independent component. This can be used additionally to the server security policy configuration to ensure that the negotiated security settings for a communication channel provide a strength required by the security policy. The independent audit option will reveal failures in the configuration of the client or server side or both.

IV. Ensuring Secure Communication

As depicted in the previous section by taking TLS as example, it is possible to monitor the security negotiation of secure communication protocols in a passive way, without interfering with the protocol. To utilize this property, an additional component – a crypto option filter – in a network is defined. This crypto filter may be realized as separate component or may be part of an already existing component handling the data exchange, e.g., a switch.

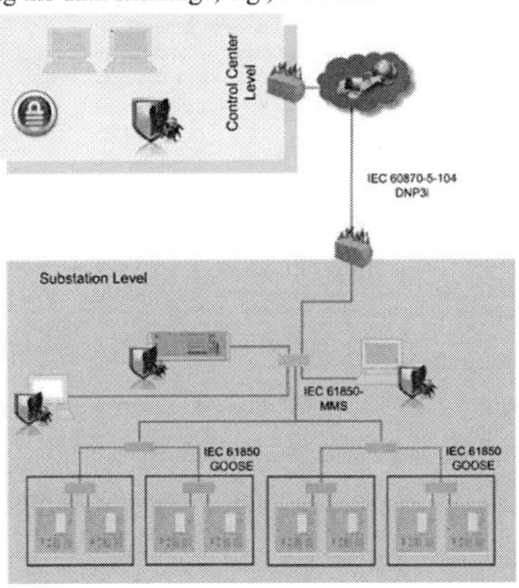

Figure 9. Substation to Control Center Communication

Figure 9 shows the underlying use case targeting the communication between a substation and a control center connected over a public network using a dedicated protocol (here: IEC 60870-5-104) for telecontrol, which is secured by TLS. Both sides are required to authenticate within TLS on the base of X.509 certificates and to provide support for one of the following cipher suites:

- TLS_RSA_WITH_AES_128_CBC_SHA
- TLS_DH_ DSS_WITH_AES_128_SHA
- TLS_DH_DSS_WITH_AES_256_SHA
- TLS_ECDHE_ECDSA_WITH_ AES_128_SHA

The following cipher suites are explicitly forbidden, as they do not provide confidentiality of the data exchange or not even integrity protection (first bullet)

- TLS_RSA_WITH_NULL_NULL
- TLS_RSA_WITH_NULL_SHA256
- TLS_ECDHE_ECDSA_WITH_NULL_SHA

This data is typically contained in a policy configuration data base together with connection specific information to identify the associated security policy.

In the following, two approaches for the realization of a crypto option filter from a network design perspective are described. This also comprises a functionality to utilize the information for ensuring a match to a given security policy, which may then lead to the interruption of communication establishment, if the security policy is not met.

Figure 10 shows a variant, in which the crypto option filter is placed directly into the communication path. This realization may be based on existing network components in the communication path. The data analysis component monitors the connection establishment and the TLS handshake without interrupting the communication channel establishment. The handshake messages *ClientHello* and *ServerHello* carry the specific information about the cipher suite negotiation, which is monitored and compared with the data from security policy database. Additionally the exchange of the server and client side certificate is monitored. As an additional service, the crypto filter may validate the exchanged certificates to ensure that they are not outdated or revoked. Depending on the match of the security negotiation parameter with the security policy, the communication establishment may be terminated through the policy enforcement component.

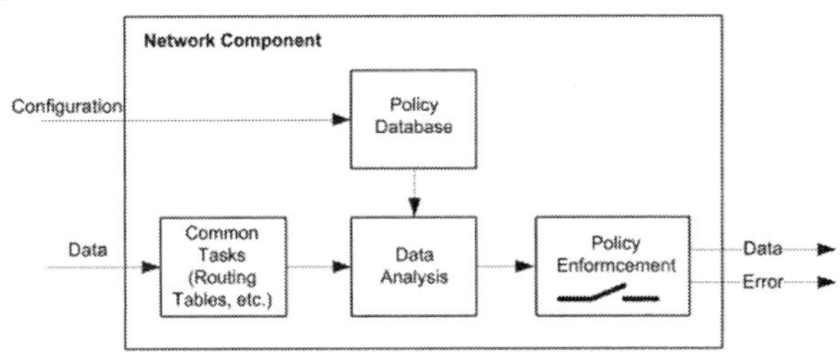

Figure 10. In-path Crypto Option Filter

In contrast to the in-path crypto option filter, Figure 11 shows an off-path filter. The general evaluation is similar to the in-path filter, with the exception of the data access. As the filter is not directly placed in the communication path, a probe on the network duplicates the traffic and forwards it to the off-path crypto option filter. This probe may be a separate component or a monitoring port on the existing infrastructure component as shown in Figure 11. If it is a separate component, the probe may already preprocess the handshake and extract the information, which can then be provided to the crypto option filter. If the functionality is included in an existing infrastructure component, the complete TLS handshake may be forwarded to the crypto option filter for inspection. Alternatively, the policy enforcement component may integrate the traffic duplication.

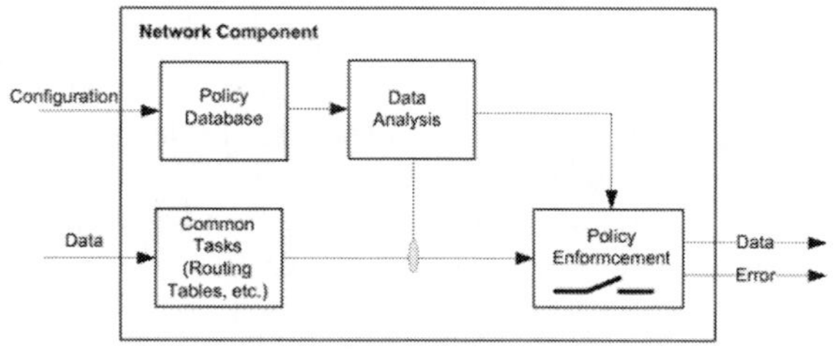

Figure 11. Off-path Crypto Option Filter

The off-path variant has the clear advantage that the policy checking component can be centralized, independent from the actual communication path to be checked.

Note that the description for the crypto option filter focused on the TLS 1.2 version as discussed in Section III. But TLS will be evolved and TLS 1.3 is currently under development. This version will result in simplifications of the meanwhile complex handshake and will reduce the available options and also shorten the handshake phase to three messages. Most importantly, TLS 1.3 will utilize the established key already in the handshake phase to protect messages. The monitoring approach described in the following is still applicable, as the message parts containing the monitoring target are still in clear.

V. Conclusions and Outlook

This paper described a solution to ensure that communication between different components of a system is in fact protected according to a dedicated security strength as defined by a given security policy. It ensures that the required level of security is indeed utilized during operation. As shown, requirements for secure communication exist through different guidelines, standards, and also legislation. The proposed solution was shown in the context of substation to control center communication, to ensure mutual authentication and an appropriate protection of the communicated information. As the smart energy grid does increasingly integrate DER systems, the chance of communicating privacy related data increases. And so do the requirements for protected communication.

The example shown related to the protocol TLS, which is used in power system automation to secure the communication. Also other protocols like IPSec or openVPN exist, which are used to provide a secure tunnel for exchanging information. Here, the initial handshake during the connection establishment can be monitored in a similar way as shown for TLS.

Moreover, as the proposed crypto filter verifies the establishment of secure communication channels according to a given security policy, it can also be used to offload further validation tasks from the communication peers, like the validation of the peer certificates utilized during connection establishment.

As stated in the beginning, this paper describes the concept for ensuring the establishment of secure communication channels. The consequent next step is the integration of the proposed approach in a prototype, to validate the effectiveness.

References

[1] T. Dierks and E. Rescorla, "The Transport Layer Security (TLS) Protocol Version 1.2", RFC 5246, Aug. 2008, http://tools.ietf.org/html/rfc5246 [retrieved: Jan. 2016].

[2] NERC-CIP, North American Electric Reliability Corporation, "CIP Critical Infrastructure Protection Standards", Version 5, http://www.nerc.com/pa/Stand/Pages/CIPStandards.aspx, [retrieved: Jan.2016]

[3] German IT Security Law, July 2015, http://www.bgbl.de/xaver/bgbl/start.xav?startbk=Bundesanzeiger_BGBl&jumpTo=bgbl115s1324.pdf (German) [retrieved: Jan. 2016]

[4] German Energy Act, EnWG, July 2012, http://www.gesetze-im-internet.de/bundesrecht/enwg_2005/gesamt.pdf (German) [retrieved: Jan. 2016]

[5] Technical Guideline TR 03109, Technische Vorgaben für intelligente Messsysteme, 2015, https://www.bsi.bund.de/DE/Publikationen/TechnischeRichtlinien/tr03109/index_htm.html (German) [retrieved: Jan 2016]

[6] ANSSI Technical Note, Recommandations de sécurité concernant l'analyse des flux HTTPS, October 2015, http://www.ssi.gouv.fr/uploads/IMG/pdf/NP_TLS_NoteTech.pdf (French) [retrieved: Jan. 2016]

[7] IEC62443-3-3:2013, "Industrial communication networks – Network and system security – Part 3-3: System security requirements and security levels", Edition 1.0, August 2013.

[8] IEC 62351-x Power systems management and associated information exchange – Data and communication security, http://www.iec.ch/smartgrid/standards/ [retrieved: Jan. 2016].

[9] ISO 27000 Series

[10] Bundesverband der Energie- und Wasserwirtschaft, Datensicherheit, BDEW "Whitepaper Requirements for Secure Control and Telecommunication Systems," Version 1.1, 03/2015., http://ldew.de/bdew.nsf/id/52929DBC7CEEED1EC125766C000588AD/$file/Whitepaper_Secure_Systems_Vedis_1.0final.pdf [retrieved: Jan 2016]

[11] NIST IR 7628 Guidelines for Smart Grid Cybersecurity: Vol. 1 - Smart Grid Cybersecurity Strategy, Architecture, and High-Level Requirements, Vol. 2 - Privacy and the Smart Grid, Vol. 3 - Supportive Analyses and References, NISTIR 7628 Rev. 1, (Volumes 1-3), http://nvlpubs.nist.gov/nistpubs/ir/2014/NIST.IR.7628r1.pdf [retrieved: Jan 2016]

Aggregators Efficiency in Distributed Power Networks

Alain Tcheukam, Hamidou Tembine

New York University Abu Dhabi, United Arab Emirates
Email: {alain.tcheukam, tembine}@nyu.edu

Abstract—**Fully distributed smart power grids approaches are increasing in the energy sector. A prosumer (producer-consumer) is a user that not only consumes electricity, but can also produce and store electricity. This paper focuses on power market models in which prosumers interact in a distributed environment during the purchase or sale of electric power. In this paper, we propose an hierarchical distributed model, which is based on reinforcement learning and optimization. Different types of prosumer are able to intelligently buy energy from, or sell it to, the power market. Our simulation results show that the integration of the aggregator in the power grid help to reduce the peak energy consumption and to lower the electricity cost for the population of prosumers.**

Keywords–distributed power networks, prosumers, aggregators, optimization, reinforcement learning.

I. INTRODUCTION

Exploratory researchs in the field of smart grid are increasing in the renewable energy sector. As energy efficiency and clean energy technologies become more common, system challenges require to rethink traditional paradigms of energy system planning and operation. A clean energy revolution is taking place worldwide. We can distinguish between two types of electricity power management systems: the centralized and the decentralized power management system [4], [5], [6], [7], [9].

The centralized power management system is currently used in many countries [1], [2], [8]. The feature of the centralized model is that at the physical layer, the grid is designed for a one-way flow of the electricity. More precisely, from the top where the electricity is generated from large power plants and transported to local substations, to the bottom which is the final stage in the delivery of electricity to end users. Moreover, at the power market layer the wholesale power market of this model can be subdivided into two category [3]: integrated (or pool) market and unbundled (or forward) market.

The idea behind the decentralized power management system is to exploit the increasing integration of decentralized energy resources (DER) into the distribution network [1], [2], [8], [12], [15], [16], [17]. DER systems are modern technologies based on solar, wind, geothermal, water, biomass, biofuel or other renewables energy resources. More precisely, they are small or mid-scale power generation technologies (typically in the range of 3KW to 10 MW). A prosumer (producer-consumer) is a user that not only consumes electricity, but can also produce and store electricity. Establishing clean or renewable energy sources involves the problem of adequate management for networked power sources. This is due to intermittency of renewable energy source during the electricity generation and to the variability of prosumer consumption/production.

In [13], the authors propose a methodology to quantify the quantity of ramping load reserves a priori. However, it is assumed therein that the probability density function of imbalances is invariant, which may not be the case of current energy systems. The work [14] studies scheduling techniques for storage devices using global information between the entities.

A. *DIPONET: a grid of micro-grids*

In this paper, we introduce a distributed power negotiation concept that enables the energy balancing in the entire power network and we term it DIPONET. Our approach is more dynamic since it provides the consumer/prosumer a more efficient way to interact in the distributed power network. The DIPONET approach is based on the previous project DEZENT [4], [5] . The limitation of the producer/consumer in the previous project is that it only make use of reinforcement learning when buying/selling energy from/to the power market. The consumer in that approach is static in the sense that it is not able to anticipate or to delay energy consumption. Moreover at the power market layer, a negative bidding strategy used by the static producer/consumer was affecting its portfolio.

TABLE I. Notations

Symbol	Meaning
$[1,2,\cdots,T]$	time horizon of an entire day
Q_t	the energy profile of an entire day. $[q_1,\cdots,q_t]$
$k \in \mathbb{Z}$	maximum energy variation
$x_t \in \mathbb{Z}$	the energy variation at time t, $-k \leq x_t \leq k$.
R_{max}	maximal energy reserve capacity $R_{max} : \mathbb{N}$
R_0	initial energy reserve capacity $R_0 : \mathbb{N}$
R_t	state of the energy reserve at time t, $R_t : \mathbb{N}$
$o : \mathbb{Z} \to \mathbb{R}$	additional cost. if $0 \leq x$ then $o(x) = x$ else $x \leq o(x) \leq 0$.
A	the the strategy space of the consumer/producer
s_t	negotiation strategy at time t, $s_t \in A$
$P(t,s)$	the value of the payoff associated to action s at time t.
$r(t,s)$	the realized payoff associated to action s at time t.
$\alpha \in \mathbb{R}$	$0 < \alpha \leq 1$
m	the number of prosumer of the aggregator

B. *Contribution*

The novelty of our approach is that it combines both distributed learning and optimization to predict supply/demand and storage in a more efficient way. More specifically we use reinforcement learning to adapt prosumer's price strategy at the power market and dynamic programming to predict future outcomes. The learning accounts for the fact that the power producer/consumer will need to use his/her best strategy when buying/selling energy quantities from/to the power market. The dynamic programming approach is due to the fact that a short term optimization might not be suitable in the event that a power consumer needs to plan his/her energy consumption over a period of a day, a week or even a month.

We show that a consumer with DIPONET scheme gets a lower cost than the one using DEZENT scheme. We then introduce the notion of aggregator in the distributed power market DIPONET. We show, through experimental study, that the aggregator is able to reduce the peak energy consumption of the grid. In our setup, an aggregator is composed of a set of prosumers. We distinguish two type of aggregator models: (i) the prosumers can share information about their expected energy price, (ii) and in the other the information is not shared. Interestingly we show that in both aggregator models the additional energy needed by the grid operator can be made available by the aggregator during the under supply period and peak load.

The rest of the paper is organized as follows. In the next section, we describe the prosumer planning methodology with both optimization and learning. In Section III we present the effect of the aggregator in the distributed power networks. Experiments and numerical results are presented in Section IV. Section V concludes the paper.

The notations used in the article are given in Table I.

II. PROSUMER PLANNING AND ADAPTATION MODEL

The power grid (see Figure 1) of interest consists of a bottom-up multi-level micro-grids architecture which is subdivided into 4 levels. The first level (0.4 kV) is a low-range network covering subdivisions (a neighborhood). The second level (10 kV) is a medium-range area network covering a suburb (regional grid). The third level (110 kV) is a long-distance energy transport network. Finally in the fourth level (380 kV) the electricity is produced from large power plants (coal, gas or nuclear). Power needs of prosumers are covered through alternative energy sources within the first 2 layers and additional power needs will be covered at the latest by the fourth level.

At the power market layer, we consider a multi-stage negotiation system through which the energy produced by micro-grids (at various tiers) are auctioned to/by the prosumers. The balancing of demand and supply between participants is carried out through balancing group managers (BGMs) which are located in different network layers and operate in parallel on each grid. The BGM balances the supply and the demand of electricity between a producer and a consumer who have submitted a similar bid.

A prosumer in such a power grid can be viewed as user who has the ability to independently modify his/her power requirements optimizing his cost (see Figure 2). In our model, we believe that independent planning by the prosumers may improve significantly the matching between production and consumption in the power grid. In practice, this could mean to help balancing the power market, since the price will favor low consumption/high production when the cost is high and vice versa. Moreover, our approach is not centralized and, in this sense, is different from demand side supply management. The idea is to exploit the (limited) ability of prosumers of planning in an autonomous way their consumption/production. Hence, they do not sign any contract leaving the planning to others: our independent consumer planning is a local matter involving only one prosumer.

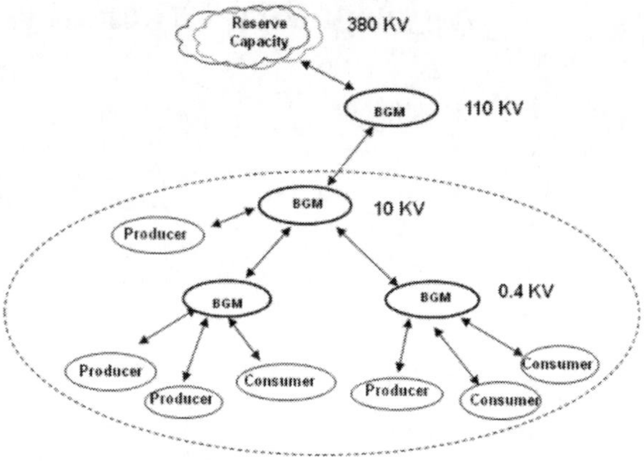

Figure 1. Power grid and associated agents [4].

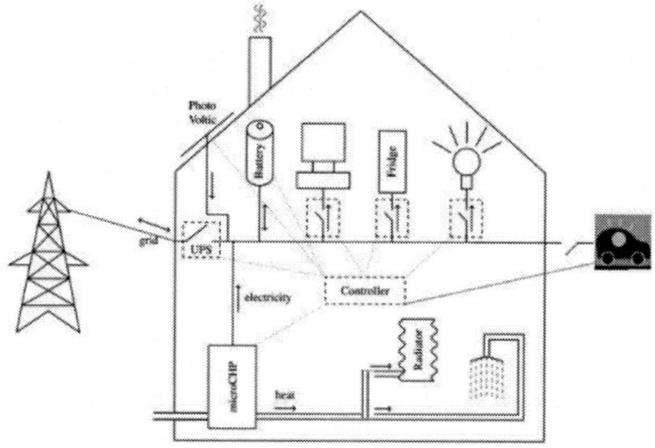

Figure 2. Model of a smart house energy consumption/production.

A. Control strategy of the Prosumer in the DIPONET approach

The challenge of the prosumer is to make elastic the demand for, and the supply of, electricity in order to optimize its energy cost based on power market conditions and on suitable constraints on their power consumption. Prosumer strategies are concerned with two separate phases. On one hand, prosumers should compete on the DIPONET power market: they should negotiate a deal with a close partner, thus achieving a better price. On the other hand, they should concentrate as much as possible their energy consumption in the periods when the prices are more convenient. For clarity we will use a consumer oriented terminology, but most of the discussion could be dualized.

We now described the combined scheme DIPONET.

Step 1: Reinforcement learning
At the end of a negotiation, the final achieved price by the consumer is normalized according to the frame size of the negotiation. Then, the temporal difference method of Sutton [10], [11] is used to derive the payoff of the negotiation strategy currently executed.

$$P(t+1,s) := P(t,s) + \alpha\,(r(t,s) - P(t,s)); \qquad (1)$$

Step 2: Optimization
For clarity we consider the case of a consumer. The consumer is characterized by the class of energy variation profiles (s)he can adopt during the day.

Controllable inputs: we denote by $x(t)$ the vector of controllable inputs of the consumer. The consumer has the ability to increase/decrease its energy consumption.

State of the consumer: the state of the consumer is given by the level of the energy reserve R_t.

Objective function: the objective of the consumer is to minimize the electricity cost achieved at the end of the day.

$$\max_{x(t),t=1,\ldots,T} \sum_{t=1}^{T}(o(x_t)+q_t)P(t,s) \tag{2}$$

Constraints
(i) summing up all the variations from the beginning of the day to any time, we cannot exceed a lower and an upper bound of the energy reserve. This constraint accounts for available energy storage media, like electric vehicle batteries or thermic accumulations due to anticipated heating, or delayed air conditioning.

$$\forall\ 1 \leq t \leq T,\ 0 \leq R_0 + \sum_{t=1}^{T} x_t \leq R_{max} \tag{3}$$

(ii) the sum of all the energy variations in the whole day for the consumer is zero, i.e. if in some slot the variation is positive, in some other slot it must be negative.

$$\sum_{t=1}^{T} x_t = 0 \tag{4}$$

Proposed algorithm: we propose an efficient dynamic programming algorithm for planning the energy consumption/production. The control algorithm of the consumer has three inputs: (i) the definition of the class of allowed profiles Q; and (ii) the cost of a unit of energy which is the result of the negotiation in each time step of the previous day. Hereafter, the optimization problem and the proposed dynamic programming algorithm used to solve it are defined.

Let $Z_t(Y_t) : \mathbb{R} \cup \{\infty\},\ t = 1,\ldots,T,\ 0 \leq Y_t \leq R_{max}$ be the optimal energy costs for time step $1,\ldots,T$, when the final energy reserve at time t is Y_t. Here $Z_t(Y_t) = \infty$ if energy reserve Y_t cannot be achieved at time step t. Thus $Z_0(Y_0)$ (no slot has elapsed yet) is everywhere ∞ except for $Z_0(R_0) = 0$.

Subproblems

$$Z_t(Y_t) = \max_{x_1,\ldots,x_t} \sum_{i=1}^{t}(o(x_i)+q_i)P(i,s)\ ,\ t = 1,\cdots,T \tag{5}$$

$$\forall i'\ 1 \leq i' \leq T,\ 0 \leq R_0 + \sum_{i=1}^{i'} x_i \leq R_{max} \tag{6}$$

$$R_0 + \sum_{i=1}^{t} x_i = y_t \quad 0 \leq y_t \leq R_{max} \tag{7}$$

Dynamic programming

$$Z_t(Y_t) = \max_{\substack{-k \leq x_t \leq k \\ 0 \leq Y_t - x_t \leq R_{max}}} Z_{t-1}(Y_t - x_t) + (o(x_t)+q_t)P(t,s) \tag{8}$$

$$Z_0(Y_0) = \text{ if } Y_0 = R_0 \text{ then } 0 \text{ else } \infty \tag{9}$$

$$Z_T(R_0) = Z \tag{10}$$

The value of Z_t at time t is computed sequentially in terms of Z_{t-1} by looking backwards for $Z_t(Y_t)$ to the optimal energy costs at slot $t-1$ for eligible values $Y_t - x_t$ of the energy reserve. Finally, an *optimal strategy* S is any sequence $S = \widehat{(\hat{x}_1, \hat{Y}_1), \ldots, (\hat{x}_T, R_0)}$ such that the values of $\hat{x}_t$ and of $\widehat{Y_{t-1}}$ are computed backwards from $\hat{Y}_t$, $t = T \ldots, 1$, by letting $\widehat{Y_T} = R_0$, the final reserve being R_0. Formally: optimal strategy

$$Z_t(\hat{Y}_t) = Z_{t-1}(\hat{y}_t - \hat{x}_t) + (o(\hat{x}_t)+q_t)P(t,s), \tag{11}$$

$$\widehat{Y_{t-1}} = \hat{Y}_t - \hat{x}_t \tag{12}$$

$$\widehat{Y_T} = R_0 \tag{13}$$

The time and space complexity of the algorithm are $O(TR_{max}k)$ and $O(TR_{max})$ respectively.

In section IV, we compare the behavior, in terms of energy cost minimization, of a consumer in our approach with that of a consumer of the DEZENT approach

III. AGGREGATOR OPTIMIZATION MODEL

We study the effect of an aggregator in the DIPONET power market. An aggregator in our model is defined as a collection of prosumers. In the proposed approach, each prosumer is neutral in the sense that it essentially neither consumes nor produces energy, as it can only sell in the power market the energy previously bought and stored. Actually, a prosumer consumes a little amount of energy, due to the overhead of the energy storing processes. Thus the behavior of the virtual prosumer is similar to that of a rechargeable battery. Only a real prosumer could combine the effect of a virtual prosumer with that of a producer and a consumer.

We define two types of aggregator: (i) a decentralized aggregator which consists of a number of prosumers running in parallel. Those agents do not exchange with each other the information about their energy achieved price. (ii) Centralized aggregator in which prosumers share (and therefore use) the same price information for the bidding and optimization phase. The objective of the aggregator in this case is to maximize its portfolio at the end of the day.

Decentralized aggregator: each virtual prosumer (of the aggregator) exploits the control model defined in section II-A.

Centralized aggregator: let m be the number of prosumer globally controlled by the aggregator.
Controllable inputs: we denote by $[x_1(t),\cdots,x_m(t)]$ the vector of controllable inputs of the aggregator. The aggregator has the ability to buy or sell energy in the power market.

Uncontrollable inputs: the uncontrollable input are the same as in the case of the costumer.

Aggregator state: $X_{Aggr}(t) = \sum_{i=1}^{m} x_i(t)$

TABLE II. Non stationary DIPONET environment setup

Architecture	Negotiation Level	1
	BGM on Level 1	1
	Clients	23
	Producers (50 – 350 KW)	10
	Consumers (50 – 300 KW)	10
	Prosumers (50 – 300 KW)	3
Electricity price	Day duration: 60 slots (24 hours)	
	Profile cost of the electricity (free market)	
prosumers environment	Producers: Gaussian distribution 1	
	Consumers: Gaussian distribution 2	
Energy reserve	finite capacity	
	prosumer initial level: 0	
Controller	Class of consumption profiles	
	Planning phase: optimization	
Simulations	Duration: 3 days	
	Test: non stationary environment	

Objective function: the objective is to maximize the the gain of the portfolio at the end of the day.

$$\max_{[x_1(t),\cdots,x_m(t)],t=1,\ldots,T} \sum_{i=1}^{m}\sum_{t=1}^{T}(o(x_t)+q_t)P(t,s) \quad (14)$$

Constraints: in our model, the allowed profiles of the aggregator must satisfy the following constraints:

(i) the energy variation in a time step for any prosumer belonging to the portfolio of the aggregator has a lower and an upper bound;

$$\forall\, i \in \text{Aggr}, \forall\, 1 \leq t \leq T,\ 0 \leq R_0 + \sum_{l=1}^{t} x_{il} \leq R_{max} \quad (15)$$

(ii) the sum of the all the energy variations in the whole day for the aggregator is zero. Notice that at the end of the day this condition must be different from zero when considering only one prosumer belonging the portfolio of the aggregator.

$$\sum_{i=1}^{m}\sum_{t=1}^{T} x_{it} = 0 \quad (16)$$

(iii) summing up all the variations from the beginning of the day to any time, we cannot exceed a lower and an upper bound. Let m be the number of prosumers of the aggregator

$$\forall\, 1 \leq t \leq T,\ 0 \leq \sum_{i=1}^{m}(R_0 + \sum_{l=1}^{t} k_{il}) \leq mR_{max} \quad (17)$$

The algorithm used by the central aggregator is similar to that of section II-A except for the fact that: (i) the achieved price used in the algorithm is the weighted average of all the prosumer of its portfolio; (ii) the constraints of the problem are both local and global and (iii) the aggregator finally allocates the energy profile for each prosumer.

IV. SIMULATION STUDIES

The setup of the space of the experiments is based on the available DIPONET and DEZENT simulator and on the implementation of the two types of aggregator. It depends essentially on three parameters: (i) the free market power cost, which can exhibit high or low variance: for this we chose real data from the day ahead market prices of Switzerland (date: March 9, 2013); (ii) the prosumers environment, namely heavy production or heavy consumption, in which the total amount of the electricity produced in the subnet is respectively greater than or less than, the total amount needed in the subnet. In the heavy consumption situations, the additional, needed power is made available at the large power plant level, at a price which depends on the time of the day. Analogously for the heavy production situations. In all these cases, the profile cost of the electricity at the global level (namely at the large power plant level) was the same for all days; (iii) the available energy reserve capacity of the virtual prosumers characterizing the aggregator is considered finite. The experiments were conducted on the NYUAD cluster at the division of Engineering laboratory, simulating a 3 days service period (see Table II). Here the last day has been considered, since in this way transitory effects are minimized.

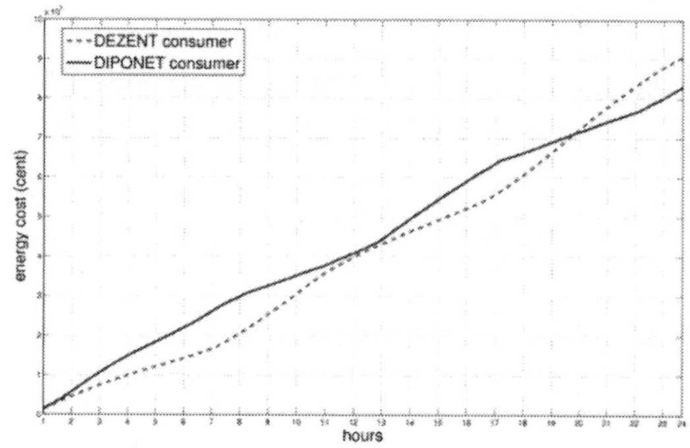

Figure 3. DIPONET vs DEZENT.

DIPONET vs DEZENT: in the first experiment we compare the behavior of a consumer in the DEZENT approach with that of our approach. The comparative study is based on the total cost of the electricity paid at the end of the last day. We recall that the consumer in DEZENT only make use of reinforcement learning while the consumer in the DIPONET approach uses both reinforcement learning and optimization. In Figure 3 the consumer in DEZENT (dashed curve) has spent more than the DIPONET consumer (solid curve). The saving of the DIPONET consumer is about 8,23%. This is due to the fact that the DIPONET consumer is able to anticipate or delay energy consumption thanks to its flexibility.

Aggregator models: in the second experiment comparative studies were based on the total cost of the electricity paid at the end of the last day by the consumer population and on the profit realized by the aggregators. Figure 4 synthesizes the optimal controller of the one prosumer of the centralized aggregator. The two upper curves of Figure 4 represent the unitary cost of energy as resulting from the negotiation phase at day 2 (solid curve) and at day 3 (dashed curve). The difference between the two upper curves gives an idea of the possible variations between the outcomes of different negotiations. Notice that the profile of the global energy cost and the context of competing prosumers is the same in both days. The lower dashed curve (respectively lower solid curve) represents the result of the optimization algorithm applied to the curve of day 2 (respectively of day 3). The curves plot the sum (from the beginning of the day) of the suggested variations: according to

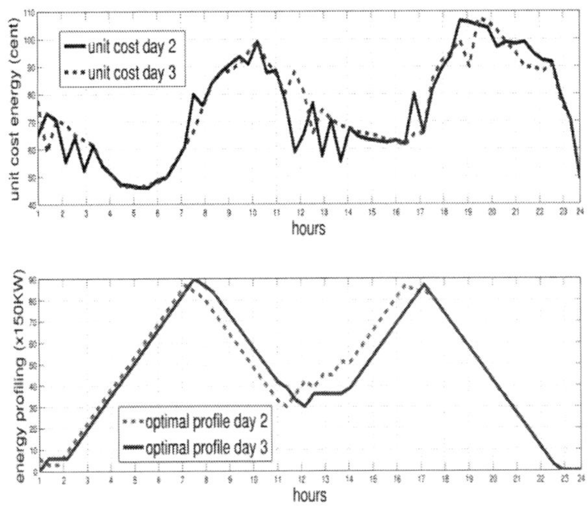

Figure 4. Optimization model of one prosumer.

the constraints we assumed on the virtual prosumers profiles, the sum of the variation must be not greater than the aggregates reserve capacity and should end up at 0. Notice that the controller correctly suggests variations which are opposite wrt. the negotiated cost.

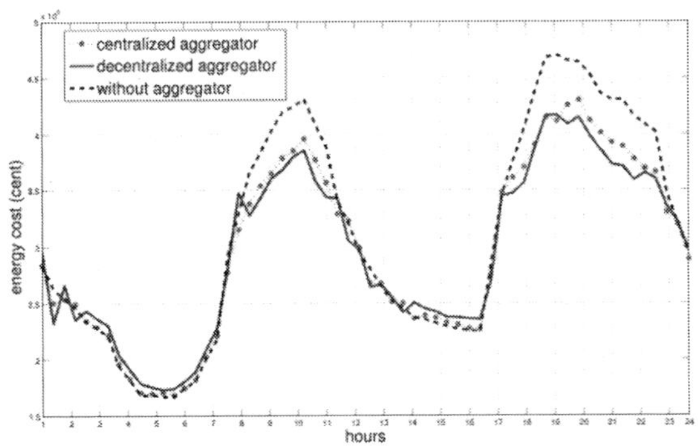

Figure 5. Energy cost achieved by the consumer population.

Figure 5 reports the result of the placebo test on the behavior of the consumer population. In Figure 5, the upper dashed curve represents the energy cost achieved in each hour by the entire population when there is no aggregator in the power market. Analogously, the middle bullet curve represents the case in which the centralized aggregator is active and the lowest solid curve the case in which the decentralized aggregator is active. The observation we have is that the consumers population when the aggregator is active has spent less during peak energy consumption period.

In Figure 6, the final cost achieved at the end of day 3 by the population of consumers in which the aggregator was active (solid curve for the decentralized approach and bullet curve for the centralized approach) is less than the case in which there

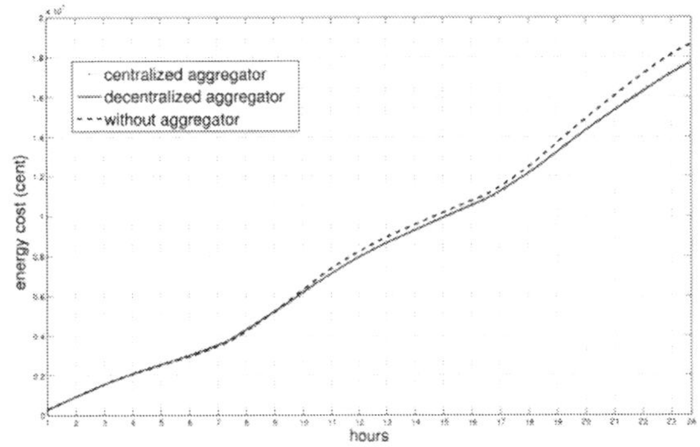

Figure 6. Energy cost at the end of the day achieved by the consumer population.

was no aggregator (dashed curve). This positive effect is due to the introduction of an aggregator. The two curves when the aggregator is active are overlapped and this is due to the fact the available reserve charactering the aggregators are the same in both cases. The percentage of global energy cost reduction is about 3,2%.

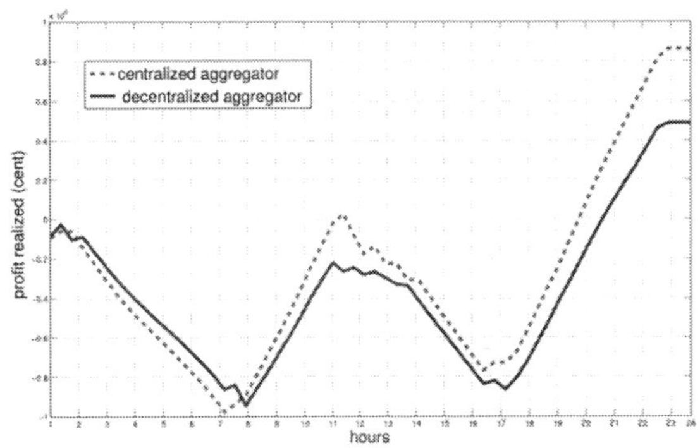

Figure 7. Aggregators: profit realized.

In Figure 7 we compare the actual profit of the two aggregator approaches. The solid curve reports the energy cost (actual profit) realized by the decentralized aggregator in the power market. The profit is given by the sum of the entire profit realized by the virtual prosumers characterizing the aggregator. Analogously, the dashed curve reports the actual profit of the centralized aggregator. The profit in the centralized case is given by the sum of energy cost achieved in every time slot. There is a remarkable difference between the two gains and the centralized aggregator exhibits a superior behavior. This is due to the fact that the centralized aggregator uses cooperative informations of virtual prosumer characterizing its portfolio. The information about the difference between the two gains of Figure 7 can be well studied if we compare the actual profit of each aggregator with its expected one.

In Figure 8, the dashed curve represents the expected profit of the decentralized aggregator at the end of day 3. The curve is computed by assuming known in advance the energy cost. That curve has been obtained by summing up all the energy

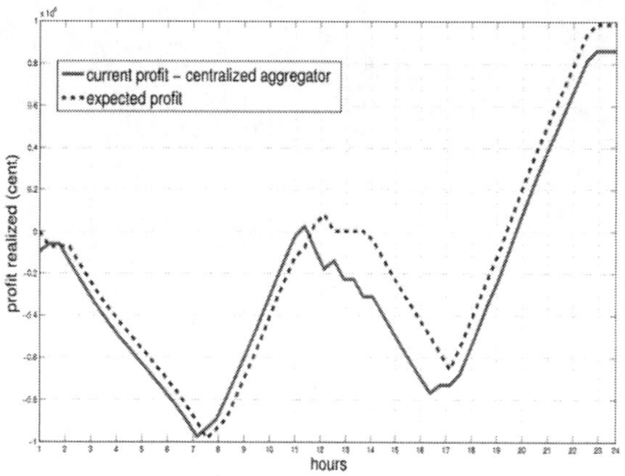

Figure 8. Centralized aggregators, expected cost vs real cost.

costs of the optimal profile of day 3 of the virtual prosumers.

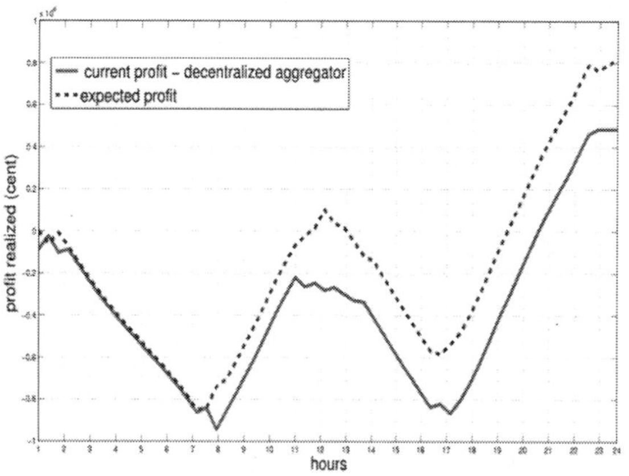

Figure 9. Decentralized aggregators, expected cost vs real cost.

Analogously, in Figure 9 the dashed curve represents the expected profit in the decentralized aggregator case at the end of day 3. That curve is given by the energy costs of the optimal profile of day 3 of the centralized aggregator. Notice that the difference between the expected gain and the actual gain of the aggregator in the decentralized case is greater than that of the centralized case.

V. Conclusion

This paper proposed the combination of reinforcement learning and optimization as a mechanism for buying/selling energy in a distributed power network. Simulations results showed that our approach is more efficient than the approach used in DEZENT. Next the paper study the effect of the aggregator on the distributed power market. The general aggregator concept is to make use of the flexibilities of the prosumers for providing active demand services in the power market. The results of the introduction of the aggregator in the distributed power grid have been the reduction of the peak energy consumption and the lowering of the electricity cost for the population of prosumers. Future works will include the development of a randomized algorithm that allow the prosumer to use different learning strategies for biding into the power market. At the optimization level, the repeated re-planning algorithm currently used by the prosumer will be extended in order to help minimize the mismatch between the anticipated energy profile and the real energy profile used during the day. Finally, the strategic interaction between prosumers will be studied through a dynamic games. Investigating on what happens if we are dealing with an infinite number of power prosumer in the distributed power grid.

References

[1] L. Barroso, T. Cavalcanti, P. Giesbertz and K. Purchala, "Classification of Electricity Market Models Worldwide," IEEE PES, International Symposium 2005, Paper 102.

[2] S. Stoft, " Power System Economics : Designing Markets for Electricity," Wiley-IEEE Press, 2002.

[3] R. Wilson, "Architecture of power market", Econometrica, 2002, pp. 1299-1340.

[4] H. Wedde, S. Lehnhoff, C. Rehtanz, and O. Krause, "Bottom-up self-organization of unpredictable demand and supply under decentralized power management," Self-Adaptive and Self-Organizing Systems, Second IEEE International Conference, 2008, pp. 74-83.

[5] H. Wedde, S. Lehnhoff, C. Rehtanz, and O.Krause,"Distributed learning strategies for collaborative agents in adaptive decentralized power systems," 15th Annual IEEE International Conference and Workshop on the Engineering of Computer Based Systems, 2008, pp. 26-35

[6] U. Montanari and A. T. Siwe, "Real time market models and prosumer profiling,"IEEE Conference on Computer Communication Workshop, pages 712, 2013.

[7] U. Montanari and A. T. Siwe, "Prosumers as Aggregators in the DEZENT Context of Regenerative Power Production," IEEE International Conference on Self-Adaptive and Self-Organizing Systems (SASO), 2014, pp. 167-174,

[8] C. Block, J. Collins and W. Ketter (2010a, August), "Agent-based competitive simulation: Exploring future retail energy markets", In Proceedings of the 12th International Conference on Electronic Commerce (ICEC), Hawaii, USA.

[9] N. Hatziargyriou, H. Asano, R. Iravani and Marnay,"An overview of ongoing research, development, and demonstration projects," in Berkeley Lab Publications LBNL-62937, Lawrence Berkeley National Laboratory, July 2007.

[10] R. Sutton, A. Barto, and C. Anderson, "Reinforcement Learning: An Introduction," The MIT Press, 1983.

[11] R. S. Sutton, "Learning to predict by the methods of temporal differences," Machine Learning, vol. 3, 1988, pp. 9-44 .

[12] A. Tcheukam and H. Tembine, "Energy cost saving strategies in distributed power networks," International Conference on Electrical Energy and Networks, Feb 18-19, 2016, Nice, France (best paper award).

[13] A. Muzhikyan, A. M. Farid and K. Youcef-Toumi, "An enhanced method for the determination of the ramping reserves," 2015 American Control Conference (ACC), Chicago, IL, 2015, pp. 994-1001.

[14] N. Rahbari-Asr, Y. Zhang, and M. -Y Chow. "Cooperative Distributed Scheduling for Storage Devices in Microgrids using Dynamic KKT Multipliers and Consensus Networks" in proceedings on IEEE PES General Meeting 2015, Denver, Colorado, USA.

[15] Alain Tcheukam, "prosumer planning in the DEZENT context of regenerative power production", PhD thesis, 2013.

[16] A. Tcheukam and H. Tembine, "Mean-Field-Type Games for Distributed Power Networks in Presence of Prosumers", the 28th Chinese Control and Decision Conference (CCDC), May 2016

[17] A. Tcheukam and H. Tembine, "Energy Cost Saving Tips in Distributed Power Networks," Book chapter 2 in Smart Grid as a Solution for Renewable and Efficient Energy, A. Ahmad, Naveed Ul Hassan (Eds.) IGI, 2016.

Optimization of Power Usage Effectiveness for Heterogenous Modular Data Centers using Neural Network

Vishal Kumar Singh
College of Engineering and Computer Science
University of Michigan-
Dearborn, USA
e-mail: vksingh@umich.edu

Dr. Jinhua Guo
College of Engineering and Computer Science
University of Michigan-
Dearborn, USA
e-mail: jinhua@umich.edu

***Abstract*- With the rise of Internet of Things (IoT), it is becoming cheaper and easier to collect data from data center (DC) mechanical, electrical and control systems. These systems have complex interactions with each other. The static control logics and high number of configuration and nonlinear interdependency create challenges in understanding and optimizing energy efficiency. This is particularly challenging and expensive in medium size or smaller configurations like data suites or modular data centers. We utilize a learning engine that learns from operationally collected data to accurately predict power usage effectiveness (PUE) and create a control model to validate test results. Using the machine learning framework developed in this paper, we are able to predict DC PUE within 0.0004 +/- 0.0005. The results show that machine learning can improve data suite efficiency. The results also indicate that neural network based controller shows promise for practical implementation.**

Keywords— Machine learning; Neural Network; PUE; Data center.

I. INTRODUCTION

Data centers are recognized as an increasingly troublesome percentage of electricity consumption in the US. A recent revision of the Koomey report [1] puts this at 2% of all US power consumption and 1.3% of worldwide power consumption. Rapid growth of cloud based systems is accelerating growth of data centers. Growing energy costs and environmental responsibility have placed the DC industry under increasing pressure to improve its operational efficiency. The development of metrics of data center efficiency (e.g., PUE) has focused attention on improving energy efficiency in data centers. Even large companies have scored low on Greenpeace report.

Figure 1. Examples of containerized/modular data center

Constructing data center space using traditional methods takes a long time. Speed of delivery of data center space has become a critical business factor for data center operators. This gave rise to modular data centers and containerized data centers. Figure 1. shows few examples of modular data center. Many companies build modular data center for inside building shell and standalone containers for outside [17] [18] [19].

Aisle containment has improved efficiency of facility side cooling power usage (chiller and fan) and load balancing of virtual servers has improved server power usage consumed by servers (IT Load) [2].

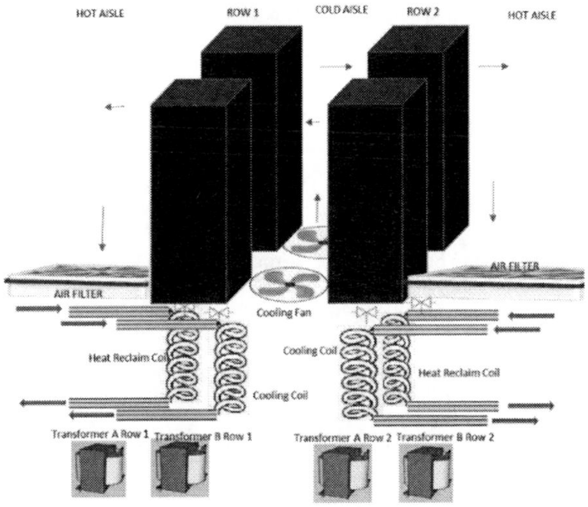

Figure 2. 2-D heterogeneous DC Experimental setup design

Internet of Things (IoT) is rapidly growing with projected $7.1 trillion by 2020 [3]. This has allowed for significant changes in asset instrumentation to communicate via internet protocol (IP). By using IoT framework, it has become possible to collect and analyze granular data from uninterruptible power supply (UPS), computer room air conditioning (CRAC), circuits, power distribution unit (PDU) etc. This allows collecting data from smaller sections of data centers like aisle, suite or data pod. Instrumenting these microsystems has allowed to manage and control smaller environment ecosystems in a data center. Figure 2. shows a modular data suite architecture and data collection points in a midsize data center.

At the given scale of power use, any incremental improvements in efficiency will produce notable cost savings and reduce carbon emissions. Large data centers like Google, Microsoft and Amazon have homogenous standard systems as compared to smaller privately held multi-tenant data center that have heterogeneous (non-standard) systems. Aisle containment and efficient virtual server load management have attempted to improve energy efficiency in data centers [3]. Metrics devised more recently like Corporate Average Data Center Efficiency (CADE) have drawn attention more broadly to all power consumption in the data center, including both cooling systems and servers, showing that there is still significant underutilization in data centers [4]. Recently, more efforts are being made to optimize data center efficiency by utilizing machine learning [5].

This paper focuses on using neural network optimizing method to predict and optimize cooling power of a given load in a modular heterogeneous data suite to optimize overall PUE.

In Section II we discuss related work. In Section III, we discuss the methodology of the neural network approach. In Section IV, we discuss results and discussion. In Section V, we discuss limitations of machine learning. Finally, we conclude in Section VI.

II. RELATED WORK

Increasing energy efficiency in a data center has been in great focus in the past few years. Efforts and have been made to optimize facilities by aisle containment [6]. There also has been work on managing virtual server loads to utilize energy efficiently [7]. There is work done in managing energy by combining building automation and virtualization together [2].

There are new demands around cloud computing, big data and infrastructure power efficiency. Furthermore, this change in the data center is being driven by more users, more data and a lot more reliance on the data center itself.

With cloud technologies and the rapid growth in data leading the way within many technological categories — working with the right data center optimization technologies has become more important than ever [9]. Data center Administrators must understand where their current energy demands are allocated and how they can best optimize those resources. Every small amount of energy efficiency gains is improvement. Recently Microsoft and Google have used machine learning techniques for energy optimization. Microsoft is measuring server workload spikes and automating data center operations [8]. Google is exploring using machine learning techniques to optimize energy use data center at a building level [5]. There has been no application of machine learning techniques in a mid-size data centers. This is due to lack of instrumenting machines and implementing IoT platform to collect and store data. The facility side infrastructure has components that have complex interactions amongst themselves. Most of the existing optimization techniques use static method such as cold aisle set point temperature. Establishing an accurate mathematical model or obtaining characteristic parameters for a proportional–integral–derivative (PID) controller in practical control scenarios is challenging, thus limiting their practical applicability [20]. On the other hand, machine learning can be accurately modeled to represent true characteristics of a DC. All the related studies for midsize data centers have been using simulations, we show results by collecting data from practical operations in mid-size data center. This study is unique in applying machine learning energy optimization technique on facility side infrastructure operational data in midsize modular data center.

This study relates to micro systems like data suites and modular data center in multi-tenant facility with heterogeneous server configurations, see Figure 2. This study is to further optimize micro facility environment related to a data suite for a given server load.

III. METHODOLOGY- MACHINE LEARNING APPROACH

Facility side infrastructure has components that have complex interactions amongst themselves. PID models do not accurately capture these interactions. Machine learning is well-suited for the DC environment given the complexity of plant operations and the abundance of existing monitoring data. The modern large-scale DC has a wide variety of mechanical and electrical equipment, along with their associated set points and control schemes. The interactions between these systems and various feedback loops make it difficult to accurately predict DC efficiency using traditional engineering formulas. We are training the neural to produce optimal set of operating parameters. Rectified Linear Units (ReLU) is used for deep learning. The model is trained to optimize for lowest PUE.

Neural Network is the machine learning approach which uses Multi-Layer Perceptron (MLP), Supervised Learning and Resilient Back Propagation Algorithm to make an efficient prediction of PUE $P_\theta(x)$ using the environmental variables n that surrounds heterogeneous DC, such as Cold Coil Temperature, Cold Aisle Temperature, Cooling Coil Chilled liquid flow, Fan Power, Chiller Power, Server Load, etc. Let us consider an x as a set of input $m \times n$, where m is the size of the dataset and n is the number of features. The input matrix is then multiplied with the model parameter θ to give the hidden layer. The size and number of hidden layers can be varied based on the complexity of the model required.

The Neural Network is adapted to DC through mathematical model framework for training DC energy efficiency models. Neural networks are a class of machine learning algorithms which adapt and react based on the behavior of neurons. They have best fit adaption, pattern searches and so on to accommodate the accuracy. The concept of machine learning is explained in detail with implementation.

A. Multi Layer Perceptron

The neural network algorithm used multi-layer perceptron, which is well applicable when modeling functional relationships. The underlying structure of an MLP is a directed graph, i.e., it consists of vertices and directed edges, in this context called neurons and synapses [10]. The neurons are organized in layers, which are usually fully connected by synapses. The synapse can only connect to subsequent layers. The input layer consists of all covariates in separate neurons and the output layer consists of the response variables. The layers in between are referred to as hidden layers, as they are not directly observable. Input layer and hidden layers include a constant neuron relating to intercept synapses, i.e. synapses that are not directly influenced by any covariate. Figure 3 gives an example of a neural network with one hidden layer that consists of three hidden neurons. This neural network models the relationship between the two covariates A, B and the response variable Y. Theoretically allows inclusion of arbitrary numbers of covariates and response variables. However, there can occur convergence difficulties using a huge number of both covariates and response variables.

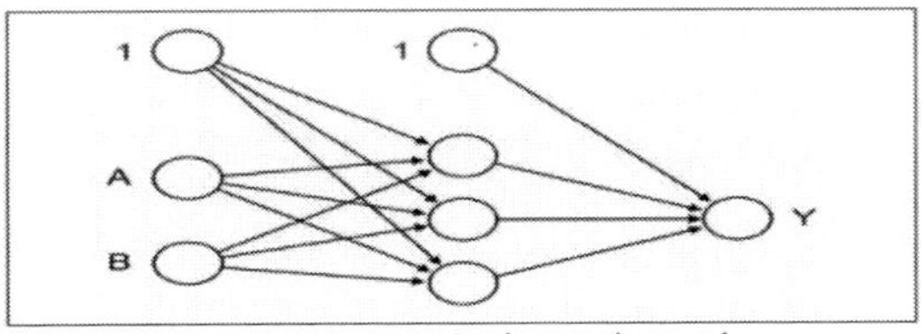

Figure 3. Example of a neural network.

To each of the synapses, a weight is attached indicating the effect of the corresponding neuron, and all data pass the neural network as signals. The signals are processed first by the so-called integration function combining all incoming signals and second by the so-called activation function transforming the output of the neuron.

The simplest multi-layer perceptron (also known as perceptron) consists of an input layer with n covariates and an output layer with one output neuron.

It calculates the function

$$o(x) = f(w_o + \sum_{i=1}^{n} w_i x_i) = f w_o + w^T x \quad (1)$$

where w_o denotes the intercept, $\mathbf{w} = (w_1,..., w_n)$ the vector consisting of all synaptic weights without the intercept, and $\mathbf{x} = (x_1,..., x_n)$ the vector of all covariates.

B. Supervised Learning

Neural networks are fitted to the data by learning algorithms during a training process which focuses on supervised learning algorithms [13]. These learning algorithms are characterized by the usage of a given output that is compared to the predicted output and by the adaptation of all parameters according to this comparison. The parameters of a neural network are its weights. All weights are usually initialized with random values drawn from a standard normal distribution.

C. Backpropagation And Resilient Backpropagation

The resilient backpropagation algorithm is based on the traditional backpropagation algorithm that modifies the weights of a neural network in order to find a local minimum of the error function [14].

D. Implementation

The machine learning algorithm used is Neural Network. The neural network utilizes 2 hidden layers and 0.01 as the regularization parameter. The training dataset contains 19 input variables and one output variable (the Suite PUE) as shown in the Figure 5b. The total size of the data samples used is 119421 rows, which were collected from a heterogeneous data center sensor ports. The 70% of the dataset is used for training with the remaining 30% used for cross-validation and testing. The chronological order of the dataset is randomly shuffled before splitting to avoid biasing the training and testing sets on newer or older data [15].

The 19 variables used for modelling are as follows.

TABLE I. SELECTED VARIABLES

Variables	Variables
Cooling Coil Leaving Temperature (°F)	Power utilized by chilled liquid (kW)
Average Cold Aisle Temperature (°F)	Cold Coil IN Water Temp (°F)
Cooling Coil Valve Position (%)	Cold Coil Out Water Temp (°F)
Fan Speed (%)	Server Load A (kVA)
Hot Aisle Temperature (°F)	Server Load A (kVAR)
Heat Reclaim Coil Leaving Temperature (°F)	Server Load A (kW)
Cooling Coil Chilled liquid flow (Gallons/Min.)	Server Load B (kVA)
Absorption (kW)	Server Load B (kVAR)
Fan Power (kW)	Sever Load B (kW)
	Suite Server Load (kW)

Data normalization, also known as feature scaling, is recommended due to the wide range of raw feature values. The values of a feature vector z are mapped to the range [-1, 1] by:

$$z_{norm} = \frac{z - mean(z)}{\max(z) - \min(z)} \quad (2)$$

The Block diagram explains the overall scenario acquired in the Data center for Predicting PUE, based on the Machine Learning Algorithm Neural Network Model.

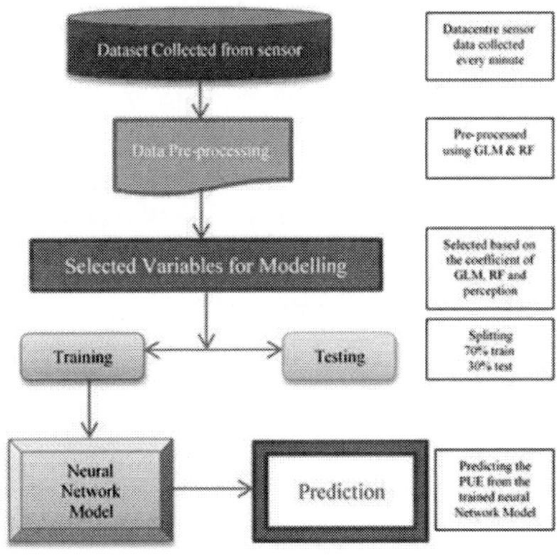

Figure 4. Block diagram of Neural Network Modelling

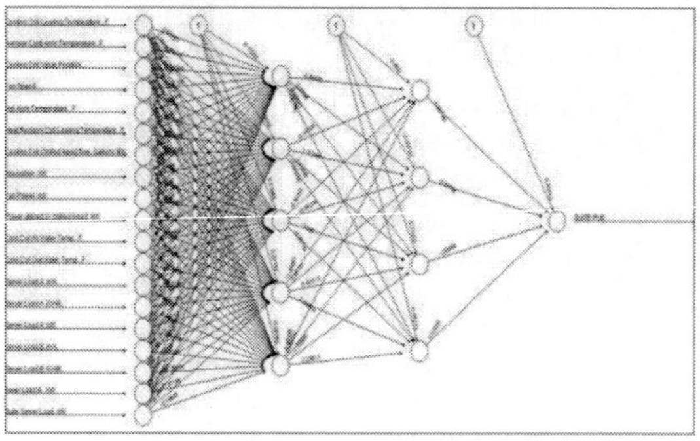

Figure 5. Network Model with selected variables

The block diagram as shown in Figure 4. represents the logic flow of neural network prediction model which evolves the processing of data retrieved from the sensor ports. The data variables are of different features which may or may not affect the SUITE PUE. The collected data is preprocessed through Generalized Linear Model (GLM) [11] and Random Forest (RF) [12] algorithm to find the effectiveness of the parameter with the coefficients. The variables are selected from the preprocessed data through positive skewness arrived with the target SUITE PUE. This achieved through the Generalized Linear Model (GLM), Random Forest (RF) and Experts Perception.

The sampling process is done for the selected variables chosen for modeling, splitting into training and testing dataset. The training data set are used to train neural network model and the testing data is used for the prediction of the data sets through the neural network trained model for the evaluation of SUITE PUE. Note that many of the inputs representing totals and averages are actually metavariables derived from individual sensor data.

Data preprocessing such as file I/O, data filtration and calculating metavariables, Variable Analysis was conducted using Excel, R. Both R and Matlab R2012a were used for model training, post processing and simulating results.

IV. Result And Discussion

The precise and robust PUE model offers many benefits for heterogeneous DC operators and owners. For example, in real time comparison of actual vs predicted heterogeneous DC performance for any given set of conditions can be used for automatic performance alerting, real-time plant efficiency assessing and troubleshooting.

A precise efficiency model also enables DC operators to evaluate PUE sensitivity to DC operational parameters. For example, an internal analysis of PUE versus Cold Aisle Temperature(°F) conducted at a heterogonous DC suggested a theoretical 0.0005 reduction in PUE by increasing the cooling tower LWT and chilled water injection pump set points by 3F. This simulated PUE reduction was subsequently verified with experimental test results after normalizing for server IT load and wet bulb temperature [5]. Such sensitivity analyses drive significant cost and carbon savings by locating and estimating the magnitude of opportunities for further PUE reductions.

Finally, a comprehensive DC efficiency model enables operators to simulate the DC operating configurations without making physical changes. Currently, it's very difficult for an operator to predict the effect of a plant configuration change on PUE prior to enacting the changes. This is due to the complexity of modern DCs, and the interactions between multiple control systems. A machine learning approach leverages the plethora of existing sensor data to develop a mathematical model that understands the relationships between operational parameters and the holistic energy efficiency. This type of simulation allows operators to virtualize the DC for the purpose of identifying optimal plant configurations while reducing the uncertainty surrounding plant changes.

A. Prediction Results

Figure 6 depicts a snapshot of predicted vs actual PUE values at one of heterogonous DCs over one month during the summer. The neural network detailed in this paper achieved a mean Square error of 0.004 and standard deviation of 0.001 on the test dataset. Note that the model error generally increases for PUE values greater than 1 .29 due to the shortage of training data corresponding to those values. The model accuracy for those PUE ranges is expected to increase over time as additional data are collected on heterogeneous DC operations.

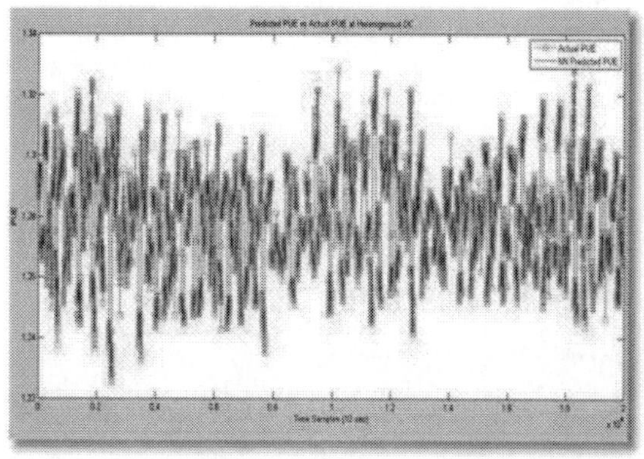

Figure 6. Predicted vs Actual PUE values at heterogonous DC

B. Sensitivity Analysis

The following graphs reveal the impact of individual operating parameters on the DC PUE. We isolate for the effects of specific variables by linearly varying one input at a time while holding all others constant. Such sensitivity analyses are used to evaluate the impact of set point changes and identify optimal set points. All test results have been verified empirically.

Figure 7a represents shows, as the Cooling coil leaving temperature (°F) increases, the PUE decreases. DC should be maintained with increasing the cooling coil leaving temperature with stabilizing other variables and making PUE more effective to reduce the cost. Similarly, Figure 7b suggests that the providing a system with cold aisle temperature (°F) over a period of time under different circumstance, the variation in the PUE is linearly increased as the cold aisle temperature decreases.

Figure 7c represents a linear variation as the Cooling coil valve position increases the PUE also increases, as it is directly proportional the usage of power is more as it becomes big. Figure 7d indicates that when Cold coil out water temperature decreases eventually the PUE increases, so the temperature for this scenario is optimized and they are inversely proportional to each other.

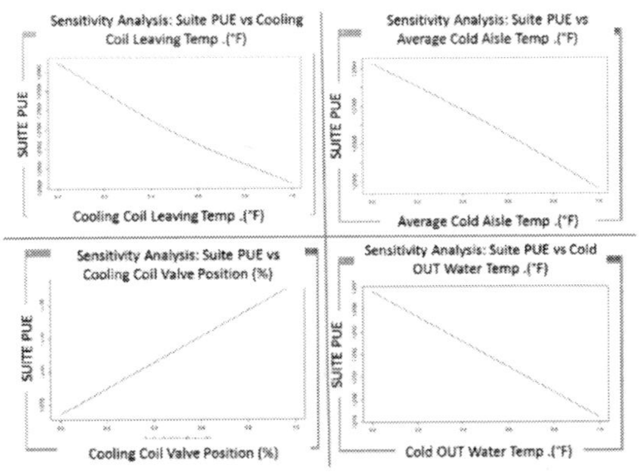

Figure 7a-7d: SUITE PUE vs Cooling Coil Leaving Temperature, Average Cold Aisle Temperature, Cooling Coil Valve Position and Cold Coil Out Water Temp

Figure 8a represents that as the cooling coil chilled liquid flow increases significantly the SUITE PUE decreases so there is an inversely proportional to each other.

Figure 8b represents a slightly sloppy curve for the SUITE PUE versus Heat reclaim coil leaving temperature (°F), says that PUE is in stabilized state when the temperature is in the optimal stage and also shows that they are inversely proportional as the temperature increases the PUE drops out.

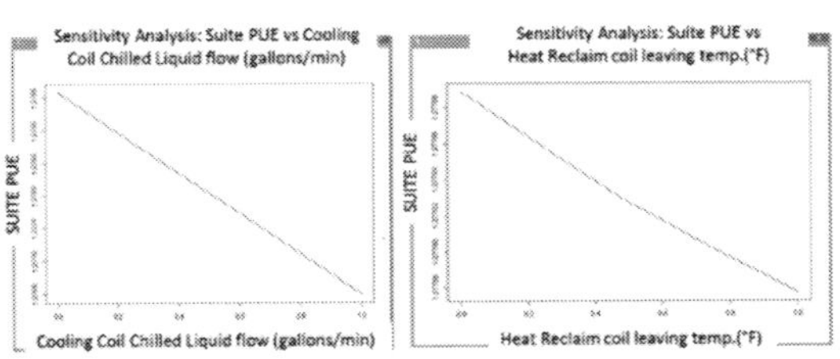

Figure 8a-8b: SUITE PUE vs Cooling Coil Chilled Liquid Flow and Heat Reclaim Coil Leaving Temperature

fan power for controlling the PUE without exceeding drastic change in the power consumption.

Figure 9a & 9b show that Fan Power and Fan Speed are directly proportional to SUITE PUE, where Figure 9a signifies a linear variation between the PUE and Fan Power but Figure 9b depicts that there is an optimization in fan speed through an upper sloppy curve which creates a positive impact in the

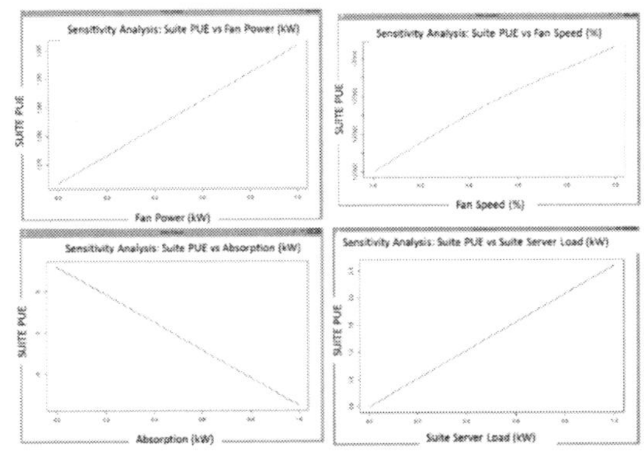

Figure 9a-9d: SUITE PUE vs Fan Power (KW), Fan Speed (KW), Absorption (KW) and Suite Server Load (KW)

Figure 9c signifies that the Absorption (KW) which is the chiller power varies inversely to PUE, as chiller power increases PUE drops. It concludes that it creates a great impact in PUE, which relatively stabilized through the fan power and server load for better synchronization.

Figure 9d specifies the variation of PUE with Suite Server Load (KW) is linear, which states that the PUE decreases exponentially as the server load decreases. Eventually as per the data samples trained most of the power in the heterogeneous DC station is consumed by server load 78%.

Figure 10 represents that the accuracy of the Neural Network model with test cases empirically verified in matlab simulation [16]. The variation of PUE from the actual calculation with Neural Network trained model gives optimized results.

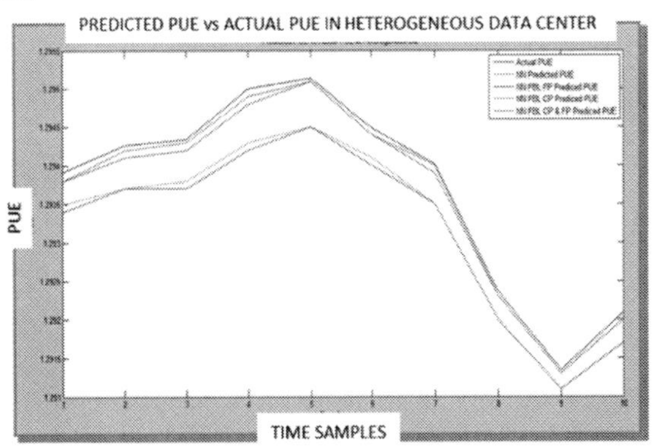

Figure 10: Neural Network Based Controller Output for DC

The model simulation done in four different scenarios for PUE optimization, such as Neural Network predicting PUE, Optimizing Fan Power (FP), which feedback to Neural network to Predict PUE, Optimizing Chiller Power (CP) which feedback to Neural Network to predict PUE and finally the best accuracy is obtained from optimizing both Fan Power (FP) and Chiller Power (CP).

Machine learning applications based on neural network based controller are limited by the quality and quantity of the data inputs. As one of important aspects to have a full range of DC operational conditions to precisely train the mathematical model. The model accuracy may decrease for conditions where there is less data. As with all empirical curves fitting, the same prediction results may be achieved for multiple model parameters θ. It is up to the analyst and DC operator to apply reasonable discretion when evaluating model predictions.

V. Conclusion

Accelerating growth in data center complexity and scale is making energy efficiency optimization increasingly important yet difficult to achieve. Though the model is simulated for heterogeneous data center environment where servers placed are of different kind, so the variation causes high end and low end rather than median. This made effective through machine learning and acquired best gain in PUE. Using the machine learning framework developed in this paper, we are able to predict DC PUE within 0.0004 +/- 0.0005. Using machine learning technique, you can further optimize power usage efficiency between 1% to 3%. This can translate is saving hundreds of thousand dollars in a datacenter. Actual testing on heterogeneous DCs indicates that machine learning is an effective method of using existing sensor data to model DC energy efficiency, and can yield significant cost savings. Model applications include DC simulation to evaluate new plant configurations, assessing energy efficiency, and identifying optimization opportunities.

Acknowledgement

I would like to thank the leadership of a midsize data center in Indiana, USA to help offer their facility of this experiment.

References

[1] Jonathan Koomey. Growth in Data centerelectricity use, 2005 to 2010. Analytics Press, Oakland, CA, 2011.

[2] Integrated Approach To Data Center Power Management Lakshmi Ganesh, Hakim Weatherspoon, Tudor Marian, Ken Birman Computer Science Department, Cornell University. 2012.

[3] IBM Interconnect 2015: A New Way. http://www.slideshare.net/ibm/ibm-interconnect-2015-asset-management-and-the-internet-of-things, May 2016

[4] http://www.data centerknowledge.com/archives/2011/11/15/pue-is-dead-the-case-for-performance-per-watt/ , May 2016

[5] Machine Learning Applications for Data Center Optimization Jim Gao, Google 2014.

[6] Impact of Hot and Cold Aisle Containment on Data Center Temperature and Efficiency Revision 2 by John Niemann, Kevin Brown, and Victor Avelar.

[7] Ramya Raghavendra, Parthasarathy Ranganathan, Vanish Talwar, Zhikui Wang, Xiaoyun Zhu." No "Power" Struggles: Coordinated Multi-level Power Management for the Data Center".

[8] Automating Data center Operations Using Machine Learning by Peter Bod.

[9] Uptime Institute. Data CenterIndustry Survey. 2013.

[10] Frauke Gunther and Stefan Fritsch," neuralnet: Training of Neural Networks," The R Journal, Vol. 2/1, June 2010, pp 30-37.

[11] C. Marschner, "glm2: Fitting Generalized Linear Models with Convergence Problems," The R Journal, Vol. 3/2, December 2011, pp 12-15.

[12] Andy Liaw and Matthew Wiener, "Classification and Regression by randomForest," The R Journal, Vol. 2/3, December 2002, pp 18-22.

[13] Derrick , "Neural Network for self Learning control system," IEEE Control System magzine, Vol. 1, April 1990, pp 18-23.

[14] Swarup, K.S. and Subash, "Neural network approach to voltage and reactive power control in power systems" Intelligent Sensing and Information Processing, Proceedings of International Conference, 2005.

[15] Freeman J. A. and Skapura D. M.,. "Neural networks: Algorithm, Applications and Programming teaching techniques" AddisonWesley.1991.

[16] Chengming Lee and Rongshun Chen," Optimal Self-Tuning PID Controller Based on Low Power Consumption for a Server Fan Cooling System," Senors, Vol 15,pp 11685- 11700,2015.

[17] Deploying and using containerized/moular data center facilities, Christopher Kelley, Cisco Systems Jud Cooley, Oracle. https://www.thegreengrid.org/~/media/WhitePapers/WP42-DeployingAndUsingContainerizedModularData centerFacilities.pdf?lang=en, May 2016

[18] http://www.networkcomputing.com/data-centers/better-data-center-standardization-through-pod-architecture-design/55918907, May 2016

[19] http://dcblox.com/about-us/data-centers-2/ May 2016

[20] Research of PID Control Algorithm Based on Neural Network Liu Luoren, Luo Jinling ESEP 2011: 9-10 December 2011, Singapore

SmartEnergyHub - A Big-Data Approach for the Optimization of Energy-Intensive Infrastructures

Florian Maier, Andreas Wohlfrom, Falko Koetter*, Sabrina Merkel-Malkowski† and Daniel Zech‡,
*Fraunhofer IAO
Nobelstr. 12, 70569 Stuttgart, Germany
Email: firstname.lastname@iao.fraunhofer.de
†Seven2one Informationssysteme GmbH
Waldstr. 41-43, 76133 Karlsruhe, Germany
Email: sabrina.malkowski@seven2one.de
‡Fichtner IT Consulting AG
Sarweystr. 3, 70191 Stuttgart, Germany
Email: daniel.zech@fit.fichtner.de

Abstract—**The Energiewende in Germany comes with growing uncertainties due to constantly changing regulatory requirements, volatile and less controllable generation from renewables and rising energy costs for consumers. As a result, operators of large-scale infrastructures, such as airports or production industries, have to reconsider their energy supply and management. Prognosis and combined optimization of multiple forms of energy are mandatory to reduce costs and to profit from monetary incentives. For solving such an optimization problem in real-time and subsequently controlling the energy supply and demand respectively, information and communication technology (ICT) is mandatory. Within the SmartEnergyHub research project an approach is developed to identify energy optimization potentials based on a big data approach, which uses high volumes of sensor data. Therefore, SmartEnergyHub combines models for energy demand prediction, as well as optimization techniques for efficient energy consumption, production and procurement to find optimal operating schedules. This paper describes the software requirements that have to be met by such a system and presents an architectural concept for its design using the Stuttgart Airport as an example.**

Keywords–energy management; big data; smart data; in-memory database; infrastructures.

I. Introduction

Infrastructure operators especially in Germany have to cope with rapidly changing energy markets and regulations as a consequence to the German Energiewende. Considerable fluctuations in electricity generated from wind and solar energy and a related increase of energy costs through the Renewable Energy Act levy force infrastructure operators to improve their energy efficiency and reinvent their energy management [1]. The energy market is currently undergoing a rapid transformation - from a consumer driven market to a producer driven market [2], which gives energy-intensive infrastructures the opportunity to actively participate in the energy market and, thus, potentially improve their economic situation.

Infrastructures such as airports, harbors, industrial or chemical parks, factories and public facilities are considerable producers and consumers of energy. They are confronted with new technologies and requirements such as energy storages, load profile forecasting or demand side management [3]. In addition to the stepwise implementation of these technologies and process models, they have to ensure that their technical infrastructure provides safe, stable and efficient energy services. In light of these developments managers of infrastructures face the challenge of ever-increasing energy costs, as well as the need to utilize the advantages offered by active energy market participation, which allows a considerable cost reduction through flexible purchasing times.

Small and medium producers and consumers of energy do usually not have the resources and knowledge to optimize their energy systems in an integrated approach. A survey of the Fraunhofer Institute stated that only 31 % of small and medium enterprises (50-249 employees) actively monitor and manage their energy supply and demand [4]. However, conventional energy management systems already reach their limits with respect to the new challenges as optimization approaches in most cases only relate to individual components. This is why innovative approaches are needed. The solution described in this paper aims at an intelligent and more efficient use of already available data by applying big data technologies. The objective is to develop a scalable energy management IT platform built on top of existing infrastructure systems in order to optimize both internal energy flows and the energy procurement strategies. The major contributions of the SmartEnergyHub project are the real-time integration of sensors in a high-performance platform, that provides a holistic optimization of the energy system, the automated computation of schedules for energy generation and consumption units, as well as methods for decision support which take the current market situation into account.

The SmartEnergyHub Architecture addresses the following challenges:

- Analysis of big streams of data needs to be done fast and reliably.
- Information related to subsystems needs to be integrated into larger contexts, which allows a comprehensive view of the system and an improvement of the energy efficiency by connecting formerly unrelated information.
- Results of the energy management optimization in the form of schedules help to adjust the system continually.
- Load forecasts need to be considered to adjust the system in advance.

- In addition to a short-term optimization long-term energy procurement strategies are taken into consideration.

This work is structured as follows. Section II gives an overview of ICT solutions, which are comparable to the SmartEnergyHub in a way that they either deal with the integration of large amounts of sensor data, offer the integration of forecasting models or enable optimization of energy systems. In addition to the introduction of state-of-the-art software platforms, several ideas, techniques and approaches in the context of smart grid architectures are explained. Section III addresses different requirements that are necessary for a sensor-based and cloud-based platform. In Section IV, the architectural concept of the SmartEnergyHub is illustrated and core processes are described.

II. Related Work

Katiraei et al. [5] provide a framework for the control and operation aspects of smart grids. They focus on existing smart grids and the description of approaches for market participation. An autonomous demand side management was developed by Mohensian-Rad et al. [6]. The model is based on a common electricity net but does not deal with heating, cooling and air condition. Agent prices are determined with game theory approaches. Real-time pricing models have advantages but are restricted to the technology (see Mohensian-Rad and Leon-Garcia [7]). There are several mathematical models that can be used for optimizing infrastructures and smart grids [8]–[10]. The security and safety of data is addressed in [11]. A communication model based on heterogenous data sources was introduced in [12]. The design of a data driven communication model was also discussed from a microeconomic point of view in [13]. Other publications focus on web-based modeling of a smart grid [14], optimal energy flow management for distributed energy hubs [15] and load file forecasting [16]–[19]. Aman et al. discussed measures for the evaluation of the prediction models [20]. The integration of fault detection and diagnosis in the context of energy consumption into a comprehensive energy management system is described in [21]. The analysis of the existing literature gives an overview of various aspects of energy management, which should be considered in the comprehensive approach in the SmartEnergyHub project. The following sections will focus on the question how these approaches can be reflected throughout the modules to build a SmartEnergyHub.

III. Requirements

In the SmartEnergyHub project, new approaches for energy optimization of large infrastructure providers are developed and applied at the Stuttgart Airport. Stuttgart Airport handles about 10 million passengers each year, which requires the operation of an infrastructure comparable with a small city. Therefore, the Stuttgart Airport takes on the role of both an energy consumer and producer. Energy generation facilities include a combined heat and power unit, photovoltaic plants, as well as emergency power generators. Energy has to be provided mainly for heating, cooling and ventilation but is also necessary for baggage handling systems or apron lighting. About 10,000 sensors distributed throughout the infrastructure area constantly record power and gas consumption, temperature or air quality and numerous other parameters, which are also partially influenced by visitor behavior. As the infrastructure and the sensor landscape is getting more and more complex, the quantity of data is growing fast. In addition, the frequency of the data collection is expected to increase. Currently most of the sensors supply values every minute or every fifteen minutes. The project SEH aims to extract necessary information about optimization potentials by using *big data processing*. Beyond the use of existing sensor data, external data sources (e.g., weather forecasts, energy price forecasts or passenger numbers) will be connected to a new scalable platform, which relies on a high performance in-memory database. The SmartEnergyHub project aims at a comprehensive approach, which takes into account all energy facilities. The main modules of the SmartEnergyHub are:

- **Data import:** The system has to support data import of different sources. These include sensor data available through the infrastructure system, weather forecasts, as well as price forward curves.
- **Data cleansing:** The system has to support the cleansing of imported raw data and offer possibilities to convert heterogeneous data to a standardized data format. One of the major challenges will be to realize a sufficient import speed and frequency in combination with a data cleaning process that ensures the necessary data quality for subsequent processes. The correct interpretation of outliers e.g., as measurement error or as an actual critical system state, plays a key role.
- **Data storage:** The system must be able to store different data types. These include time series of sensors, weather and price data but also master data, calculation results and system control plans.
- **Optimization:** The system must generate system control plans for the next 24-hours with regard to a minimization of energy costs. Changes in key parameters should trigger a recalculation or adaption of the control plans.
- **Market interaction:** The system must create recommendations for optimizing the energy procurement process to support a procurement manager to make efficient decisions.
- **Forecasts:** The system should be able to forecast the energy consumption based on historic and current sensor, weather data and further parameters. Those forecasts are a necessary input for the optimization and the market functions. Whereas the optimization has its focus on the near future and thus needs short-term predictions for the upcoming hours, market functions require a long-term forecast of load curves for the next 2 to 3 years.
- **User interface:** The system must support decisions, as well as offer near-time visualization. This includes also the approval of system control plans through authorized users with the help of a user interface. The system must also present energy procurement recommendations to the user.
- **Control plans:** The system has to be able to pass control plans to the infrastructure system after approval through authorized users.

IV. Architecture

Based on the functional and non-functional requirements described in the previous section, an architecture was designed.

To ensure a sufficient level of performance, the in-memory database SAP HANA [22] was chosen as the application backbone. Database operations needed during the generation of optimal control plans, forecasts or energy procurement recommendations, are encapsulated in an intermediate layer called *Model-Interface*, which exposes common operations via a generic interface supporting interchangeability of individual modules.

A. *Modular Architecture*

The system has a modular design as illustrated by the architecture diagram in Figure 1. Modules, displayed as boxes in the figure, expose their functionality via interfaces and communicate with other modules with the help of a message broker. The architecture contains a data, application and a presentation tier. General purpose functions are provided by the module 'Basic Services and Orchestration'. The system architecture consists of the following modules:

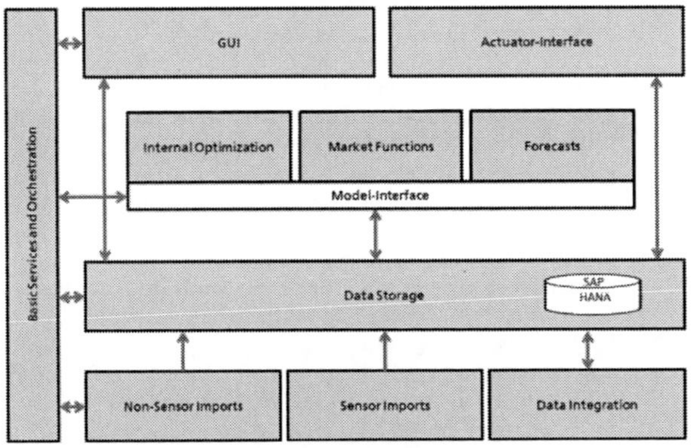

Figure 1. Overview of architecture and methodology (focus of this work highlighted in grey).

- **Sensor-Imports:** Sensor-data is imported through an interface to the existing infrastructure system, which is a building control system in the case of the Stuttgart Airport. Imported raw data is converted into a standardized data format.
- **Non-Sensor Imports:** Non-sensor data in this case is data that is not collected by sensors within the infrastructure, but is offered by other providers (containing data like weather or energy prices). This data is also imported and then converted to the standardized data format and stored in the database.
- **Data Integration:** This module transforms imported raw data into adjusted data using a generic time series data model for incoming sensor data and a hierarchical data model to store sensor master data.
- **Data Storage:** This module provides functions related to the storage and processing data within the database.
- **Internal Optimization:** Based on sensor, master and forecast data an optimal infrastructure control plan for energy facilities is generated. Database operations are executed with the help of the model interface, which offers generic methods for database interactions. The optimal infrastructure control plan is written to the database.
- **Market Functions:** Based on sensor and forecast data this module creates action recommendations for interaction with energy markets. Those action recommendations are also written to the database. Conceived as a decision support system for the energy manager, this module should also allow a smooth handover of chosen interaction strategies
- **Forecasts:** Based on sensor and external forecast data, this module generates forecasts, which can be used by the Internal Optimization module or the Market Functions module. Thus the forecast data is written to the database.
- **GUI:** The GUI module provides a user interface to support especially the approval process of generated infrastructure control plans or action recommendations related to the energy market.
- **Actuator Interface:** After a new infrastructure control plan has been created and approved by an administrator, the control plan is passed to the infrastructure system.
- **Basic Services and Orchestration:** This module coordinates the module communication and sequence control.

B. *Core Processes*

The following core processes provide a dynamic cross-module view of the planned system:

1) Import: The import process is responsible for loading all relevant data into the database. This includes sensor data which is available through the infrastructure system like sensor or master data, as well as data from external sources. The received data is transformed into a standardized data format and saved into the database. After recording the data in the database, subsequent processes can be triggered via the message broker.

2) Data Cleansing: After the recognition of new data the data cleansing process starts and writes modified data back to the database. Data cleansing supports for example the interpolation of missing values.

3) Forecasts: Within the forecast process all relevant data such as sensor, weather and energy price data is used to generate forecasts for the expected load curve. For accessing the input data, the model interface is used, which offers generic methods to read and write data, but also to process data within the database in case of time-critical operations. After a new forecast curve was produced this data is written to the database.

4) Internal Optimization: During the internal optimization process, an optimization algorithm is used to generate an optimal infrastructure control plan for energy relevant facilities. The input for the optimization algorithm consists of sensor, weather and energy price data, as well as calculated forecasts. The optimization algorithm will consider both technical and contractual constraints. For example, reaction times of technical systems, as well as legal requirements concerning clean air policies, will be modeled. Generic methods of the model interface can be used for both loading data from the database, as well as for delegating calculations to the database. Upon completion of the algorithm, the result in form of an optimal infrastructure control plan is written to the database.

5) Market Functions: The process market functions supports the energy procurement and the interaction with energy markets through recommendations. Sensor, energy price data and forecasts serve as input. Based on both current and predicted energy prices as well as the energy load curve, an optimal procurement plan is developed.

V. Preliminary Evaluation

The requirements described in this paper can be met by the presented system architecture. The system contains several modules, which support the required functions such as importing data from different sources, data cleansing, generating optimal control plans, forecasts and procurement recommendations, as well as human approval and passing control plans to the infrastructure system. Building on top of a *in-memory database*, the system supports high performance sensor data processing. The requirements were collected in cooperation with the Stuttgart Airport as the first pilot user. The evaluation of the system will start in 2016 as a pilot project and help to verify and ensure the feasibility of the developed solution. Through a continuous operation of the system until the end of the project 2017 the components will gradually be improved and extended. Further infrastructure managers and users who will be involved in a newly founded user group, will take part in the implementation and evaluation process as well.

VI. Conclusions and Outlook

This paper describes the current state of the SmartEnergyHub architecture, which lays the foundation for a high-performance smart data platform to optimize energy management in infrastructures such as airports, factories and public facilities. Applying big data technologies to energy management systems, SmartEnergyHub helps to make use of all kinds of available sensor data to optimize internal and external energy flows. In addition to meter and sensor data, external data providers are connected to the platform, which allows to take additional information such as weather or energy price forecasts into account. The data pool, which is enriched in this way, enables the generation of accurate short and long term load forecasts with regard to the overall system, as well as for subsystems. The forecasts itself are used to optimize the system in an integrated approach to control the individual plants and components proactively. This also allows to use current market opportunities such as temporary low-price phases to reduce energy costs. The required core processes and corresponding modules (see Figure 1) are described in detail in this paper. In future work, the implementation of the modules as well as the enhancement of the system architecture is planed. One of the major challenges for the future implementation consists in the design of forecast and optimization algorithms for huge amounts of data so that they can meet the real-time requirements. By involving further infrastructure operators the transferability of the solution will be shown.

Acknowledgment

The work published in this article was funded by the Bundesministerium für Wirtschaft und Energie (BMWi) under the promotional reference 01MD15011C, www.smart-energy-hub.de). The SmartEnergyHub project is a joint work of: Fichtner IT Consulting AG, Flughafen Stuttgart GmbH, Faunhofer IAIS, Faunhofer IAO, in-integrierte informationssysteme GmbH and Seven2one Informationssysteme GmbH.

References

[1] E. Gawel et al., "The future of the energy transition in germany," Energy, Sustainability and Society, vol. 4, no. 1, 2014, pp. 1–9.

[2] German Federal Ministry for Economic Affairs and Energy, "An electricity market for germanys energy transition (green paper)," 2014.

[3] Y. Simmhan et al., "Cloud-based software platform for big data analytics in smart grids," Computing in Science & Engineering, vol. 15, no. 4, 2013, pp. 38–47.

[4] R. Jochem, P. Karcher, and M. Siemer, "Betriebliches Energiemanagement in produzierenden Unternehmen Deutschland 2013," Fraunhofer IPK, Tech. Rep. 4, 2013.

[5] F. Katiraei, R. Iravani, N. Hatziargyriou, and A. Dimeas, "Microgrids management," Power and Energy Magazine, IEEE, vol. 6, no. 3, 2008, pp. 54–65.

[6] A.-H. Mohsenian-Rad, V. W. Wong, J. Jatskevich, R. Schober, and A. Leon-Garcia, "Autonomous demand-side management based on game-theoretic energy consumption scheduling for the future smart grid," Smart Grid, IEEE Transactions on, vol. 1, no. 3, 2010, pp. 320–331.

[7] A.-H. Mohsenian-Rad and A. Leon-Garcia, "Optimal residential load control with price prediction in real-time electricity pricing environments," Smart Grid, IEEE Transactions on, vol. 1, no. 2, 2010, pp. 120–133.

[8] M. C. Bozchalui, S. A. Hashmi, H. Hassen, C. A. Cañizares, and K. Bhattacharya, "Optimal operation of residential energy hubs in smart grids," Smart Grid, IEEE Transactions on, vol. 3, no. 4, 2012, pp. 1755–1766.

[9] A. Parisio, C. Del Vecchio, and G. Velotto, "Robust optimization of operations in energy hub," in Decision and Control and European Control Conference (CDC-ECC), 2011 50th IEEE Conference on. IEEE, 2011, pp. 4943–4948.

[10] P. Mancarella and G. Chicco, "Real-time demand response from energy shifting in distributed multi-generation," Smart Grid, IEEE Transactions on, vol. 4, no. 4, 2013, pp. 1928–1938.

[11] S. Uludag, S. Zeadally, and M. Badra, "Techniques, taxonomy, and challenges of privacy protection in the smart grid," in Privacy in a Digital, Networked World. Springer, 2015, pp. 343–390.

[12] S. Uludag, K.-S. Lui, W. Ren, and K. Nahrstedt, "Secure and scalable data collection with time minimization in the smart grid," Smart Grid, IEEE Transactions on, vol. PP, no. 99, 2015, pp. 1–1.

[13] M. Steurer, M. Miller, U. Fahl, and K. Hufendiek, "Enabling demand side integration–assessment of appropriate information and communication technology infrastructures, their costs and possible impacts on the electricity system," E. SmartER Europe Smart Energy Research at the crossroads of Engineering and C. Science, Eds., 2015.

[14] A. Vaccaro, M. Popov, D. Villacci, and V. Terzija, "An integrated framework for smart microgrids modeling, monitoring, control, communication, and verification," Proceedings of the IEEE, vol. 99, no. 1, 2011, pp. 119–132.

[15] M. La Scala, A. Vaccaro, and A. Zobaa, "A goal programming methodology for multiobjective optimization of distributed energy hubs operation," Applied Thermal Engineering, vol. 71, no. 2, 2014, pp. 658–666.

[16] N. Amjady, F. Keynia, and H. Zareipour, "Short-term load forecast of microgrids by a new bilevel prediction strategy," Smart Grid, IEEE Transactions on, vol. 1, no. 3, 2010, pp. 286–294.

[17] G. Zhang, B. E. Patuwo, and M. Y. Hu, "Forecasting with artificial neural networks:: The state of the art," International journal of forecasting, vol. 14, no. 1, 1998, pp. 35–62.

[18] G. P. Zhang, "Time series forecasting using a hybrid arima and neural network model," Neurocomputing, vol. 50, 2003, pp. 159–175.

[19] R. Sevlian and R. Rajagopal, "Short term electricity load forecasting on varying levels of aggregation," arXiv preprint arXiv:1404.0058, 2014.

[20] S. Aman, Y. Simmhan, and V. K. Prasanna, "Holistic measures for evaluating prediction models in smart grids," Knowledge and Data Engineering, IEEE Transactions on, vol. 27, no. 2, 2015, pp. 475–488.

[21] L. M. Blanes, A. Costa, N. Réhault, and M. M. Keane, "Integration of fault detection and diagnosis with energy management standard iso 50001 and operations and maitenance of hvac systems," Clima 2013, 2013.

[22] SAP. Sap hana. [Online]. Available: http://hana.sap.com (04, 2016)

Study of a Smart Energy Community PV and Storage Requirements A Modeling Approach Towards Net-Zero Energy

Emilio J. Palacios-Garcia, Antonio Moreno-Munoz,
Isabel Santiago, Isabel M. Moreno-Garcia, and Rafael J. Real-Calvo

Department of Computer Architecture, Electronics and Electronic Technology
University of Cordoba
Cordoba, Spain
Email: {p92pagae, amoreno, el1sachi, p92mogai, rafael.real}@uco.es

Abstract—**The paradigm of smart energy community refers to a set of households that share a Microgrid and have local renewable production and distributed energy storage. In this context, the main issue is the low dispatchability of the renewable generation, which requires large storage capacities to maximize the degree of self-consumption. A more favorable scenario is found when considering a grid connected system, which reduces the dependence on the stochastic nature of renewable resources, allowing the consumers to import energy from the grid when neither local production nor storage is available. Moreover, the surplus generation could also be injected into the electrical network, having in some cases feed-in tariffs. Thus, this paper aims to present a simulation scenario where the interaction between a Smart Community with high penetration of Photovoltaic generation and the main grid is studied, and how this integration can drive to a net-zero energy system. Results for self-consumption and self-generation indexes are presented for different Photovoltaic power rates and storage capacities, showing that net-zero smart energy communities are a plausible scenario when bidirectional energy flows with the grid are considered.**

Keywords–smart energy community; solar power generation; energy storage; load modeling; net-zero energy buildings.

I. Introduction

Among all the energy demand, the residential sector represents an average 30% of the total energy consumption in most of the developed countries, showing an upward trend in the last 10-years period with an average growth of about 5% in the European Countries. In parallel with the consumption, the electricity production also presented an average 30% increment from 1990 to 2010, but with a clear decrement of the energy produced from solid fuels (18.8%) and a higher penetration of new renewable sources whose annual average growth rate is currently 7% [1].

However, the strong mutual dependence between the renewable sources production and the weather conditions reduces the operational flexibility and the dispatchability of these generations at utility-scale, and usually requires complex forecasting models and high temporal resolution data in order to assure and planning an uninterruptible power supply within the quality standard requirements [2].

Due to the limitations of renewable resources, the classical concept of a centralized grid is moving forward to a distributed scheme, where the solar, wind or hydro production is not tied to the grid, but to the load or the consumers, creating small subsystems usually referred as Microgrids or Smart Communities. These systems have more control capabilities than a large network, can integrate different types of distributed energy resources (DER), as well as energy storage systems (ESS), and can be either connected or disconnected from the main grid [3].

Nevertheless, the evaluation of the production and storage requirements is the main issue in these networks and it needs a previous knowledge of the electricity demanded by the Microgrid or Smart Community and the available production along the year. Moreover, in the case of DER, both production and demand must be known with a sufficient resolution, especially in solar power system, where the maximum production peak and the maximum power demand are hardly ever coincident [4]. These models, together with the usage of historical data of DER, constitute the main tools for the estimation of the system requirements.

In this field, high-resolution demand modeling techniques have shown the ability to generate detailed consumption profiles. These techniques do not only take into account the seasonality of the consumption, but also predict the daily variations which cause the non-time coincidence between production and demand. In addition, some of these models allow testing future scenarios or energy policies due to their high flexibility [5][6][7].

In addition, the impact of different DER penetration rates has to be quantified in the context of load matching, meaning the evaluation of the interplay between the DER, the consumers' demand and the availability of ESS. In this field, previous work has proposed a set of indexes that study this relationship with different temporal resolutions [8][9].

In this paper, the benefits of high-resolution modeling techniques and the usage of real production data will be addressed in the context of a smart energy community to achieve a net-zero energy interaction during 1 year, but using 1-minute resolution data, so the variations in production and consumption are taken into account.

The paper is structured as following. In Section II, the methodology of the simulation model is discussed, exposing the different parts that made up the simulation system. Section III will provide and discuss the result obtained for the simulation scenario. Finally, Section IV is dedicated to the conclusions obtained from the study and the future works.

II. METHODOLOGY

In order to study the interaction between the DER, the ESS, the consumers' consumption and the grid, a simulation framework was developed. For this aim, a community composed of 200 households is simulated which share a common low voltage network and a unique feeder and whose different blocks are represented in Figure 1.

Each household is considered to have different AC loads, as well as on–site Photovoltaic (PV) generation and a local home energy storage system (HESS). The PV panels are connected through an unidirectional DC/AC converter to the household network, whereas the home ESS converter is bi-directional. In addition, each home has a bi-directional Smart Electricity Meter to quantify the energy interchanged with the grid.

On the top of this conceptual definition, the DER and HESS of each household are controlled by a centralized Smart Community Energy Management System (SCEMS). The different blocks are explained in the following sections.

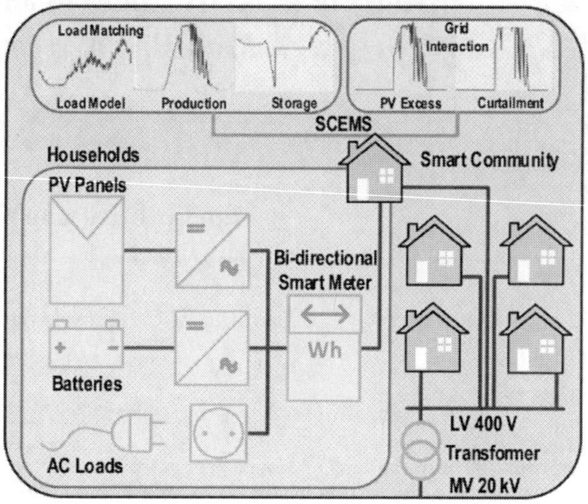

Figure 1. Conceptual Schema of the Smart Community being modeled.

A. Demand Model

The consumption profiles of each home were obtained using a stochastic model based on Markov-Chains probability theory and Monte-Carlo techniques. The model is composed of four algorithmic blocks that estimate the daily occupancy profiles, the lighting system demand, the consumption of home appliances and the energy needs of cooling and heating equipment respectively. The model was developed using the JAVA high-level programming language, and the simulations are requested and accessed through a RESTFul API.

The residents' behavior is the common influence factor in determining the energy consumption, as it has been shown by previous work [10]. Therefore, the lower level block was the one responsible for calculating the daily occupancy profile for a given household. The algorithm has a 10-minutes resolution, and its input parameters comprise the number of residents, the location and the type of day (weekday or weekend).

The occupancy model is based on non-homogeneous Markov-Chains. Therefore, for each 10-minutes instant (144 in a day), the probability transition matrices were calculated. For this aim, the Time Use Survey (TUS) was employed [11]. In this survey, carried out in most European countries, the interviewees wrote down in a diary information about the activities performed during the day, where they took place and whether they were accompanied.

Above this block, and using the generated occupancy profiles as an input dataset, the power demand was calculated applying the three other blocks, each one with specific influence factors and with a 1-minute resolution.

The lighting demand block is influenced by the solar irradiance profile, which determines the hours of the day where it is more likely to have electricity consumption due to the home lighting system. In addition, the lighting spots of the household are randomly selected, based on a probability distribution of lighting technologies and powers. As in the case of the occupancy model, the lighting consumption block has been already validated and published by the authors [5].

In the case of the appliances consumption block, besides the occupancy profile, the daily probability for different activities was taken into account. These activities are: doing the laundry, ironing, cleaning the house, watching the TV, washing and dressing, cooking, and using the PC, and their probability distributions were extracted from TUS. The usage of each home appliance is linked to one of those activities.

Finally, the block that estimates the cooling and heating consumption uses the daily ambient temperature profiles and the annual seasonality as influence factors. In this way, the usage of either cooling or heating systems could be determined during the year, as well as the hours of the day with the higher probability of consuming energy.

B. Production Data

The photovoltaic production was emulated using monitored data, obtained from historical registers of a rooftop PV installation located in Cordova (Cordoba), Spain, during 3 years. The system studied is composed of 3 similar sectors. Each sector has 36 solar modules with a peak power of 165 W, which results in a total peak power of 5,940 W per sector.

At the same time, the sectors are associated with an inverter of 5,000 W. The inverters are capable of monitoring parameters such as the input DC voltage and DC current, global irradiance, output AC voltage and AC current, frequency and power, all of them with a 5-minutes resolution.

The output AC power of each inverter was used for the simulation since it already takes into account both the losses and the efficiency of the system. This variable was linearly scaled using the quotient between the selected PV peak per household and the original one of the installation.

In addition, due to the different temporal resolutions of the historical production data (5-minutes) and the consumption model (1-minute), the output power of the inverter was linearly interpolated. This approximation is fully justified, since the power fluctuations that might be produced by the clouds are much slower.

C. Battery Storage

As it was indicated in Figure 1, each household is also considered to have an attached HESS. For our study, a simplified battery model was selected, whose basic operation is ruled by (1), which represents the charge and discharge process, and whose operative range is denoted in (2).

$$E_B(t) = E_B(t-1) + T \cdot P_B(t) \tag{1}$$

$$E_{B_{min}} < E_B(t) < E_{B_{max}} \tag{2}$$

In this equation, t represents the simulation time in minutes, $E_B(t)$ is the stored energy for each simulation step in Wh, P_B is the instant power applied to or supplied by the storage system in W, and T is the simulation step, selected to be 1 minute, but since E_B is in Wh its value is $1/60$. Regarding (2), $E_{B_{min}}$ is the maximum discharge level sometimes referred as a percentage of the maximum capacity and $E_{B_{max}}$ is the maximum charge threshold.

Both $E_{B_{max}}$ and $E_{B_{min}}$ are the main influence parameters in the simulation, since they will determine the effective capacity of the system. Different technologies will differ in the maximum deep of discharge that can be applied ($E_{B_{min}}$). Moreover, the maximum capacity of the system ($E_{B_{max}}$) could be varied along the years to simulate the aging effect, although in this study, it has not been considered.

It should also be pointed out, that although the battery system seems to be extremely simple, it allow us to represents the basic operation of the ESS as an energy reservoir. Nevertheless, current efforts are focused on the improvement of this model, so performance ratios of different technologies, as well as operation procedures, are included in the model.

D. Evaluation Indexes

The performance of DER for different values of PV power and storage capacities for de HESS were analyzed with a set dimensionless indicators. Two indicators were selected in order to study the variations in the supply utilization, the percentage of self-consumption and the amount of energy that can be injected into the grid.

The first index was the demand cover factor (DCF), which evaluates the percentage of demand that can be supplied by the PV power installed. The second was the supply cover factor (SCF) that indicates the percentage of utilization of the local generation. These two indicators are well defined in the literature, although different names are given for them [9][8].

Their expressions are indicated in (3) and (4) respectively, where $P_{PV}(t)$ is the instantaneous PV production, $P_B(t)$ is the instantaneous power supplied by ($P_B > 0$) or applied to ($P_B < 0$) the battery, and $P_D(t)$ is the power demand. Therefore, these indexes can be calculated for different periods of time, but using the instantaneous power, so the non-temporal coincidence between production and demand is considered.

$$DCF = \frac{\sum \min\left[P_{PV}(t) + P_B(t), P_D(t)\right]}{\sum P_D(t)} \tag{3}$$

$$SCF = \frac{\sum \min\left[P_{PV}(t) + P_B(t), P_D(t)\right]}{\sum (P_{PV}(t) + P_B(t))} \tag{4}$$

E. SCEMS Strategy

The last block of the simulation is the SCEMS, which controls the power interchange between the different units in the system that are the consumer loads, the DER, the HESS and the grid. The interaction defined between the system elements is indicated in Figure 2. It was implemented using MATLAB-SIMULINK environment, so the production data are loaded from a developed database, and the consumption profile obtained using the above-mentioned RESTFul service of the demand consumption model.

$P_{net}(t) = P_{PV}(t) - P_D(t)$
$E_B(t) = E_B(t-1) + T \cdot P_{net}(t)$
if $E_B(t) > E_{B_{max}}$ **then**
  $P_G(t) = E_{B_{max}} - E_B(t)$
  $E_B(t) = E_{B_{max}}$
else if $E_B(t) < E_{B_{min}}$ **then**
  $P_G(t) = E_{B_{min}} - E_B(t)$
  $E_B(t) = E_{B_{min}}$
else
  $P_G(t) = 0$
end if
$P_B(t) = -[P_{net}(t) + P_G(t)] = P_D(t) - P_{PV}(t) - P_G(t)$

Figure 2. Pseudocode of the interaction between the PV generation, the demand, the storage system and the main grid.

The control algorithm aims to maximize the autonomy of the Smart Community, controlling the energy stored in the HESS denoted as $E_B(t)$. Therefore, the distributed storage is charged as soon as the different between PV production and consumption P_{net} is positive, ($P_{net} > 0$) in order to accumulate energy for the non-production period.

This process can continue until the batteries reach their full charge ($E_{B_{max}}$), in this moment, the smart community start to interact with the grid and the production surplus that can not be stored is injected into the grid ($P_G < 0$).

On the other hand, when the available PV production is too low or zero ($P_{net} < 0$), the consumers' demand is supplied by the batteries until they reach the lower operative threshold ($E_{B_{min}}$), in this moment, the energy demanded by the community is imported from the main grid ($P_G > 0$).

After considering the above-exposed cases, the SCEMS strategy determines the instantaneous power supplied by ($P_B > 0$) or applied to ($P_B < 0$) the storage system, that will be such that the net power sum equals zero without taking into account additional losses in the system.

III. Results

Following the previously exposed methodology, a simulation scenario is presented, where the influence of the PV power and the storage capacity on the above-mentioned indexes, according to the SCEMS strategy, in the context of a 200 households community is studied.

The obtained results are shown in Figure 3, where the X-Axis represents the installed PV power peak per household, whereas the Y-Axis indicates either the annual DCF (solid lines) or the annual SCF (dashed lines), calculated with 1-minute resolution during a whole year, using the real production data, and the demand profiles simulated with the exposed model. In addition, the different results for a set of HESS capacities per house are illustrated in different colors.

Figure 3 depicts that DCF (solid lines) increases when the PV power peak increases, whereas the SCF (dashed lines) the opposite trend is observed. In contrast, for a given PV power peak, if the capacity of the HESS is increased both the DCF and the SCF are improved, but not by the same percentage.

Further conclusions can be extracted if the interaction between the DCF and the SCF is analyzed. Both indexes intersect for a given battery capacity when the PV power peak is variated. In these points, the SCF and DCF have the same value, which means that the percentage of demand that cannot be covered with the DER and the HESS equals the generation that can be neither consumed nor stored.

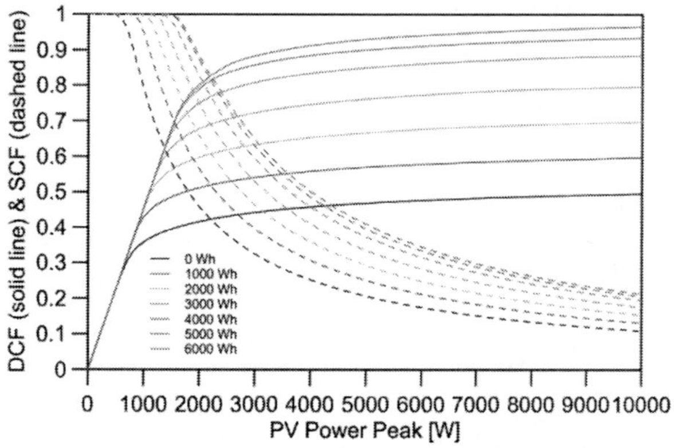

Figure 3. SCF (dashed lines) and DCF (solid lines) variation with the installed PV power and the storage capacity.

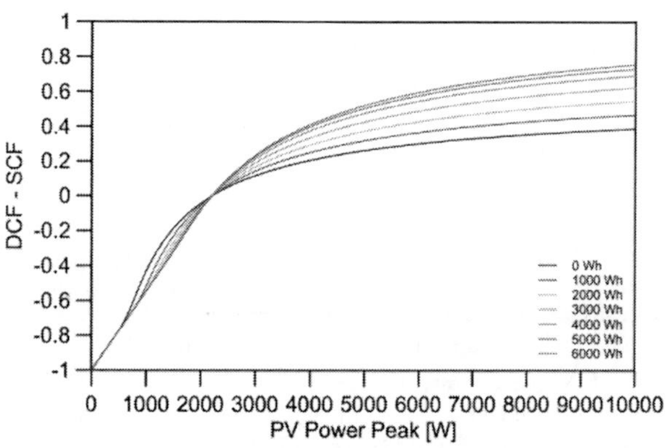

Figure 4. Difference between DCF and SCF.

However, when the system is considered grid-connected, the excess of generation $(1 - SCF)$ could be injected into the grid, whereas the demand that cannot be covered $(1 - DCF)$ must be supplied from the main grid. Hence, in this point, the exported and imported energy are similar, and the Smart Community would achieve an annual net-zero energy.

Further conclusions can be extracted if the difference of these both indexes is considered. As illustrated in Figure 4, DCF, and SCF intersect always for a similar PV power peaks independently of the battery capacity. This can be proved if (3) is divided by (4) and the both terms are multiplied by the time period T, so the power expressions can be transformed into energy units.

$$\frac{DCF}{SCF} = \frac{\sum P_D(t) \cdot T}{\sum (P_{PV}(t) + P_B(t)) \cdot T} = \frac{E_D}{E_{PV} + E_B} \quad (5)$$

In (5), the DCF equals the SCF if the consumed energy during the day E_D is similar to the energy produced E_{PV} plus the energy exchanged with the storage system E_B. Nevertheless, if the battery is considered to have a cycle per day, the net energy exchanged with the storage system is null and consequently, the intersection of both indexes only depend on the match between generation and consumption.

From (5), it could also be seen that in an ideal context, no battery storage will be necessary to achieve a net-zero energy consumption throughout the year. Nevertheless, in this case, all the produced energy that can be used should be injected to the grid, which in some cases might produce an overload in the existing networks toward the idea of hosting capacity.

If this is achieved, the net energy interchange with the electrical grid will be zero along the year, providing the grid has an infinite capacity, whereas the storage capacity will determine the additional level of the on-site generation that could be later consumed.

IV. Conclusion and Future Work

This paper has presented a comprehensible simulation scenario where the benefits of high temporal resolution models and historical data in the estimation of DER power and HESS requirements of a Smart Community have been shown by means of a set of selected cover indexes, The results have illustrated the possibility of achieving net-zero energy communities if a grid interaction is considered.

More detailed simulation scenarios will be addressed in following works, including parameters, such as the battery performance or the influence of economic factors like feed-in tariffs or variable energy prices. In addition, the system is currently being analyzed in the context of Hosting Capacity, which determines the maximum amount of PV power that can be installed without affecting the quality and reliability of the supply. Regarding this topic, future papers will be published.

Acknowledgment

This work is supported by the Spanish Ministry of Economy and Competitiveness under Research Project SCEMS, TEC2013–47316–C3–1–P.

References

[1] Eurostat, "Energy, transport and environment indicators," Publications Office of the European Union, Luxembourg, Tech. Rep., 2015.

[2] J. Widén, E. Wäckelgård, J. Paatero, and P. Lund, "Impacts of different data averaging times on statistical analysis of distributed domestic photovoltaic systems," Solar Energy, vol. 84, no. 3, March 2010, pp. 492–500.

[3] P. Palensky and D. Dietrich, "Demand Side Management: Demand Response, Intelligent Energy Systems, and Smart Loads," IEEE Transactions on Industrial Informatics, vol. 7, no. 3, August 2011, pp. 381–388.

[4] S. Cao and K. Sirén, "Impact of simulation time-resolution on the matching of PV production and household electric demand," Applied Energy, vol. 128, September 2014, pp. 192–208.

[5] E. J. Palacios-Garcia et al., "Stochastic model for lighting's electricity consumption in the residential sector. Impact of energy saving actions," Energy and Buildings, vol. 89, February 2015, pp. 245–259.

[6] J. Widén and E. Wäckelgård, "A high-resolution stochastic model of domestic activity patterns and electricity demand," Applied Energy, vol. 87, no. 6, June 2010, pp. 1880–1892.

[7] I. Richardson, M. Thomson, D. Infield, and C. Clifford, "Domestic electricity use: A high-resolution energy demand model," Energy and Buildings, vol. 42, no. 10, October 2010, pp. 1878–1887.

[8] I. Sartori, A. Napolitano, and K. Voss, "Net zero energy buildings: A consistent definition framework," Energy and Buildings, vol. 48, May 2012, pp. 220–232.

[9] J. Salom, A. J. Marszal, J. Widén, J. Candanedo, and K. B. Lindberg, "Analysis of load match and grid interaction indicators in net zero energy buildings with simulated and monitored data," Applied Energy, vol. 136, December 2014, pp. 119–131.

[10] M. Lopez, I. Santiago, D. Trillo-Montero, J. Torriti, and A. Moreno-Munoz, "Analysis and modeling of active occupancy of the residential sector in Spain: An indicator of residential electricity consumption," Energy Policy, vol. 62, November 2013, pp. 742–751.

[11] National Statistics Institute of Spain. Ministry of Economy and Competitiveness, "Time Use Survey," Spain, Tech. Rep., 2010.

Finding Needles in a Haystack: Line Event Detection on Smart Grid PMU Data Streams

Duc Nguyen, Scott Wallace, Xinghui Zhao

School of Engineering and Computer Science
Washington State University
Vancouver, WA USA
Email: {duc.nguyen, wallaces, x.zhao}@wsu.edu

Abstract—**Smart Grid technology, in particular Phasor Measurement Units (PMUs) provide a mechanism for monitoring the state of a power system across a wide geographical area at high resolution and with high fidelity. These measurements form a large corpus of state information that power systems engineers and researchers can use to find and analyze interesting phenomena either post-hoc, or in real-time. In this paper, we present our work with machine learning to develop an event detector for use with the Bonneville Power Administration's (BPA's) current PMU installation. Our system can be used post-hoc or in real-time and focuses on identifying line faults in the data stream since these events can be easily verified by records BPA maintains. One challenge for machine learning algorithms is that the modern transmission systems are very often well-behaved. Since PMUs record measurements at 60 samples/sec and each PMU typically records up to 16 phasor signals, each PMU records upwards of 80 million measurements per day. Line events, in contrast, happen very rarely (on the order of 100 per month), at least on a transmission system such as BPA's. In this paper, we examine the performance of multiple classifiers within this power system domain. In addition to examining classifier performance on a set of validated line events, we perform a detailed analysis of false alarms and explore multiple methods for reducing false alarms in a real system.**

Keywords–Smart Grid; Phasor Measurement Unit (PMU); Machine Learning; Event Detection

I. Introduction

Smart-grid technologies offer the potential to increase efficiency and reliability of power transmission by facilitating precise monitoring and control over power demand and availability. Phasor measurement units (PMUs) are a key technology for monitoring the health and status of a power transmission system. These devices record system state variables (e.g., voltage and current phase angle and magnitude among others) at high data rates (typically 60 samples/sec). Critically, PMUs also provide precise timestamps on their measurements by using the Global Positioning System. This allows the variables measured across a wide geographical area to be synchronized.

Currently, PMUs are used primarily for monitoring purposes. However, in the future it is expected that they will play a much larger role in real-time operations and control. Some of these functions may include monitoring transmission lines for faults that may occur when lines contact one another or come into contact with a grounded object. Such occurrences typically result in an observable sag in voltage at nearby substations and may require relays to actuate to isolate the affected transmission line from the rest of the grid.

In this paper, we present recent work applying machine learning techniques to detect line faults. Our experiments are conducted on archival data from a large, active power transmission system spanning the Pacific Northwest region of the United States of America. The main contributions we present include: (1) a cascade classifier that improves upon both hand-built classification rules provided by a domain expert and upon a corresponding decision tree inferred by Weka's [1] J48 implementation of the learning algorithm C4.5 [2]; and (2) a detailed examination of the classifier's error rates and an evaluation of methods for reducing false positives (normal grid activity classified as a fault).

In the following sections, we begin with a brief review of related work using machine learning for fault detection in the grid. In Section III, we describe our initial efforts to detect line faults with machine learning. These approaches serve as a baseline for the contributions described in this paper. In Section IV, we describe our dataset and then follow in the subsequent section with a discussion of two new cascade classifiers and their relative performance. In Section VI, we focus our examination on the false positive rate and on ensuring that the tradeoff between true positives and false positive stays within reasonable limits. Finally, in Section VII, we examine the cascade on a significantly larger volume of data from normal operation (nearly 100 million examples) and compare the cascade's false positive rate to that of the baseline decision-tree classifier.

II. Related Work

The increasing number of PMUs, as well as their high sample rates, present challenges for the traditional workflow of data processing [3]. As machine learning techniques have become the *de facto* standard for processing and analyzing large amount of data in other scientific fields, a variety of these techniques have also been applied to analyze PMU data for the purpose of recognizing synchrophasor signal patterns or signatures of events. In general, these approaches can be divided into two categories: unsupervised learning and supervised learning. The widely adopted techniques in these two categories are clustering and classification, respectively. The former is well suitable for identifying events with unknown signatures and the latter is particularly useful for classifying events based on a known taxonomy. Antoine et al. propose a clustering method based on PMU measurements to identify causes for inter-area oscillations [4]. In [5], the hierarchical clustering method is applied to identify dynamic event locations. The hierarchical clustering method has also been used to analyze disturbance events [6]. Clustering methods with unsupervised learning do not require labeled training data. However, once the clusters are generated, an expert's knowledge is often needed to interpret and compare the clusters [4].

In contrast to clustering, supervised learning approaches [7] can take advantage of labelled training data and identify known signatures or patterns without domain knowledge from experts. These approaches have also been used to detect interesting events from PMU data streams. For instance, Zhang et al. propose a classification method for finding fault locations based on pattern recognition [8]. The key idea is to distinguish a class from irrelevant data elements using linear discriminant analysis. The classification is carried out based on two types of features: nodal voltage, and negative sequence voltage. Similar classification techniques are used to detect voltage collapse [9] and disturbances [10] in power systems. Specifically, Diao et al. develop and train a decision tree using PMU data to assess voltage security [9]. Ray et al. built Support Vector Machines and decision tree classifiers based on a set of optimal features selected using a genetic algorithm [10]. Support Vector Machine-based classifiers can also be used to identify fault locations [11] and to predict post-fault transient stability [12].

Although significant effort has been focused on mining PMU data for detecting event signatures, very few projects use real PMU data collected from a wide area monitoring system. To the best of our knowledge, there is no previous work focusing on reducing false positives in classifying PMU data. Because of the stability of the power grid, line faults are rare phenomena. As a result, reducing false positives of the classification becomes critical. In the work presented in this paper, we examine real PMU data collected from the BPA's operational grid and propose several methods to reduce the number of false positives produced by supervised classification.

III. Baseline Fault Detection

In 2014, we described a simple set of decision rules for identifying line faults from their voltage sags [13]. These rules were developed using a theoretical foundation and validated on faults recorded over an 11 month period on the Bonneville Power Administration's transmission grid which covers a large geographic area in the Pacific Northwest of the United States of America.

Although the expert-built classification rules performed well (Table I, first row), they require a tremendous amount of effort to generate. Indeed, hundreds of pages of plots and data were generated for roughly 110 candidate faults. Clearly, the manual approach does not scale well, and as a result we have pursued machine learning approaches to leverage the data recorded by PMUs before and during these faults.

In our initial approach, we were able to show that the J48 decision tree learner was able to vastly improve upon the domain expert's hand crafted rules by: (1) leveraging a substantially larger amount of training data; and (2) improving upon the classic features directly measured by PMUs. We selected decision tree learning for these experiments because the learned representation is directly comparable to the expert's rules. Thus, we can identify how scaling up the data volume and changing the feature set contribute to performance independent of influences caused by changing the learned representation of the decision surface. While both the expert generated rules and the inferred decision trees perform classification well at PMUs closest to where the faults occur, the learned decision trees perform much better when used at an arbitrary PMU location (hence significantly higher recall values for J48 than the expert defined rules).

Table I illustrates these results. Boolean precision and recall refer to the boolean classification task of determining whether a particular moment in time should be considered a "line-event", or "normal operation". Accuracy is measured across all four possible classifications (three line-event classes: single line to ground fault (SLG); line-to-line fault (LL); and three-phase fault (3P); and the normal operating condition (N)). Macro precision and macro recall are macro-averaged values over the three line event types. We exclude the precision/recall figures from normal operating condition in the marco average since performance on normal data is best evaluated by the boolean performance characteristics. Table I illustrates a notable boost in recall for both the boolean and individual fault types cases (column two and five respectively). Moreover, while we can see that increasing the data (row 2) used for training the classifier improves recall, using a modified feature set also substantially improves the macro-recall of individual fault classes (column five).

TABLE I. Initial Performance of J48 vs Hand-Coded Rules

	Boolean Precision	Boolean Recall	Accuracy	Macro Precision	Macro Recall
Expert Rules	100	34.4	77.2	96.8	20.9
J48 Increased Data	97.2	97.8	96.5	78.2	53.0
J48 Increased Data + New Features	97.2	97.8	97.6	77.5	91.1

While the results do indeed show that both J48 classifiers make substantial improvements in recall, there is a tradeoff in precision. Most notably, boolean precision drops from 100% on the test-set for the expert defined rules to 97.2% for both learned classifiers. At one level, this seems like a reasonable tradeoff for the gains that are made on other metrics. However, a loss of almost 3% precision (column 1) means many false alarms; in a real operating environment, that value may be prohibitively high.

The remainder of this paper focuses on alternate classifiers and techniques that address the potential problem of false alarms in the J48 classifier listed above. We begin by discussing the evaluation dataset.

IV. PMU Dataset

For this study, we leveraged archival PMU data recorded by BPA between October 2012 and September 2013. We extracted examples of normal operation and each of the three fault types from within this sample. BPA verified 100 examples of line-events from within our sample using their operational database.

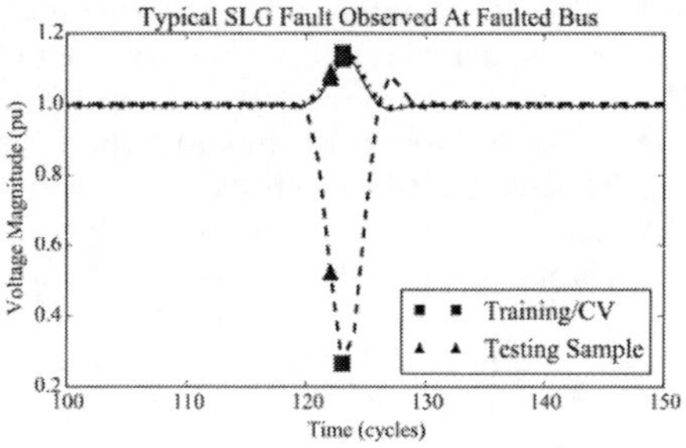

Figure 1. Typical voltage signature during a line event

Fifty-seven of these included log notes sufficient to determine a specific fault type (e.g., single-line-to-ground, line-to-line, or three-phase). The remaining faults were plotted and manually inspected to verify their voltage signatures and infer their type.

The voltage magnitude signature of a typical fault is illustrated in Figure 1. For any given fault, the PMUs reporting the largest voltage magnitude deviations from their steady state values tend to be physically close to the location of the fault. In contrast, PMUs located geographically distant from a fault may experience a minor deviation from steady state that is not easily discernible from normal variance. Unless otherwise indicated, the fault instances used to train and test the SVM classifiers described below use samples obtained across *all* PMUs with valid signals, regardless of the magnitude of the deviation experienced by each PMU. The procedure for obtaining these fault samples is as follows:

For each of the 100 fault events we found the moment in time where the measured voltage sagged to its lowest relative value. For most faults, PMUs report this moment as within 1 or 2 cycles (1/60 seconds) of each other. The most commonly reported time, for each fault and across all PMUs, was labeled the "moment of maximum voltage deviation" (square point in Figure 1).

Given the moment of maximum voltage deviation for each fault, we then collected examples using all PMU signals that measured three valid voltage phases at those moments in time. In total, the approach yields 7125 SLG examples, 474 LL examples, and 267 3P examples. To this set of 7688 fault instances, we added 11419 examples of normal operation randomly selected from across 800 minutes where BPA reported no line events occurred. Together, these samples comprised the training data for our Support Vector Machine (SVM) classifiers.

We created a test-set using the same 100 fault events but sampling them 1 cycle (1/60 second) earlier than in the training data (triangular point in Figure 1). To this, we added an additional 11419 samples of normal data obtained in the same manner as the normal data from the training set. Although the fault samples in our test set are not entirely independent of those in the training data, we believe that these points serve as a reasonable proxy for new archival data with unseen faults. Indeed, they may even be a more challenging test than new events since the selection methodology ensures that they will typically be smaller deviations from steady state than the points used in training.

Each example in the training and testing set consists of the three-phase voltage measurements for a single point in time, from a single PMU. Voltages are normalized by steady-state values calculated over a short window (10-30 cycles) that precedes each example. The voltage within the window is subjected to a smoothness check prior to computing the steady state; if this test fails, the window is moved backward in time relative to the sampling point until a suitably smooth voltage can be found as reference. In Figure 1, for example, when obtaining a measurement for the moment of maximum voltage deviation (square points at Time=122 cycles), the steady state window would initially be located so as to include cycle 121 since this point immediately precedes the measurement. However, cycle 121 deviates significantly from the mean value over the preceding 10 cycles. Thus, the smoothness criteria will not be satisfied until the window is backed up one cycle earlier so that it ends at cycle 120. This approach aims to compute a reasonable steady state value that is not adversely influenced by the voltage deviations experienced near a fault. We applied this technique to normalize all voltage measurements for both the training and testing data described above.

Using this dataset, we compare the historically best performing decision tree (row 3 in Table I) against classifiers built using SVMs in Section V. We then explore how our classifiers stack up when asked to classify a day of continuous data from 19 operational PMUs.

V. Cascades of SVMs

The main benefit of J48 is that its classification rules can be directly compared to expert rules and validated by power systems engineers. By itself, this is a substantial value. However, we anticipated that other learning algorithms may be able to improve upon J48's performance. To this end, we examined using two-class SVMs [14] in three different configurations: a standard "one-against-one" approach for multi-class learning; configured in a 3-Stage SVM Cascade (illustrated in Figure 2); and configured in a 5-Stage SVM Cascade (illustrated in Figure 3).

The one-against-one method is the default approach for learning multi-class problems with SVMs in Python's scikit-learn version 0.16 [15], which we leverage for our work. In our domain, there are four classes (three fault classes and the normal operation class). The one-against-one method learns six binary classifiers covering the $\binom{n}{2}$ class pairs. The classifiers then vote on the final prediction.

The 3-Stage and 5-Stage SVM Cascades, in contrast, learn individual binary SVMs representing a single decision node. Thus, the cascade acts similarly to a decision tree in that a series of SVMs which make progressive refinements to the set of potential classifications until a single classification is obtained. Unlike the one-against-one SVM which uses a single set of hyper-parameters for all binary SVMs that are learned, we allow each node in the Cascade to have independently set hyper-parameters which we obtained with a grid search over a subset of the training data.

The 3-Stage SVM Cascade was explored because it replicates the decision order performed by earlier J48 classifiers. The 5-Stage SVM Cascade was explored because we observed that making the fault/no-fault classification last, rather than first, may generate fewer false alarms. This is because different fault/no-fault decision boundaries could be found once the primary characteristics had a chance to group the examples into their most likely constituent fault type.

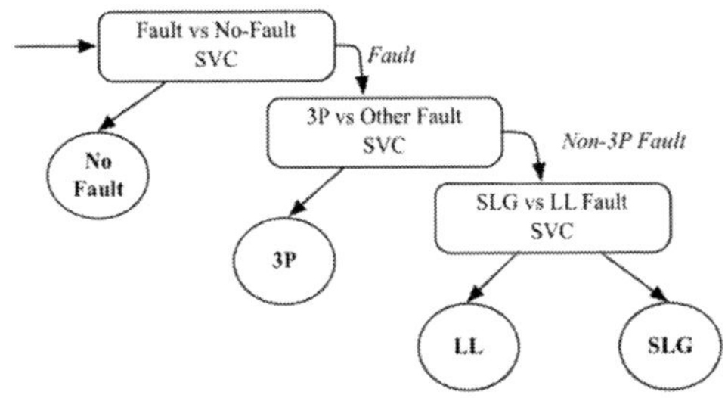

Figure 2. 3-Stage SVM Cascade

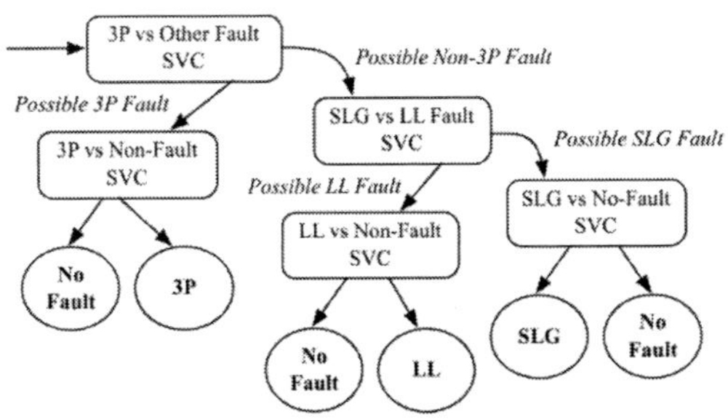

Figure 3. 5-Stage SVM Cascade

Confusion matrices for the original J48 classifier on the new test set, along with the confusion matrices for the three SVM variants are illustrated in Tables II and III respectively.

The J48 Classifier's performance, illustrated in Table II, shows high accuracy (95.5%), but a false positive rate of 4.9% (normal data classified as a fault: $(96 + 167 + 294)/11419$) that will be likely be unacceptable in a continuous operation deployment.

TABLE II. TESTING SET PERFORMANCE OF J48 CLASSIFICATION

True Classification \ Predicted Class	SLG	LL	3P	NF
SLG	6935	37	7	146
LL	124	348	0	2
3P	1	0	266	0
NF	96	167	294	10862

The SVM classifier performance is illustrated in Table III. The One-vs-One configuration (left figure) and the 3-Stage SVM Cascade (middle figure) show relatively similar accuracy (94.3% and 92.9% respectively) as the original J48 classifier and likewise similar false positive rates (4.5% for both configurations). The 5-Stage SVM Cascade, in contrast, maintains an accuracy similar to the 3-Stage Classifier (93.0%), but reduces false positives to 1.7%, nearly a factor of 3 below all other classifiers examined.

VI. TOWARDS FULL-TIME OPERATION

The results from Section V illustrate that each of the various classifiers makes somewhat different tradeoffs in terms of accuracy and false positive/false negative rates. Before selecting one classifier for further study, however, we wanted to explore how the false positive rate was impacted by modifications to the training data.

Recall that a PMU measuring a nearby line event tends to produce a relatively large deviation from steady state voltage. In contrast, PMUs located relatively far from a fault location may measure a voltage deviation that is small enough to be nearly lost in the normal fluctuations of the system. Thus, one approach to reducing the false positive rate would be simply to ignore measurements from PMUs that are geographically distant from a fault location. Intuitively, this should not be problematic from a detection standpoint since our dataset contains a large number of PMUs over a wide geographical area (covering two states in the Pacific Northwest of the United States of America). Thus, removing a portion of the PMU measurements should still provide enough measurements to classify the event.

Figure 4 shows how each of the four classifier's false positive rate changes as a result of thresholding the fault instances. Specifically, the figure plots a representative point for each of the four classifier's performance as represented in the confusion matrices in Tables II and III. For the J48 classifier whose decision nodes are discussed in [13], we change the threshold on the first decision node which corresponds to the fault/no-fault distinction; we do not retrain the classifier as a whole. As the threshold is tightened, the net effect is to classify examples with small deviations (which tend to occur far from a fault location) as normal. We plot the impact of this modified threshold as a line (blue dashed line). Note that the true positive rate reported on the Y-axis is exactly that: the true positive rate on the *unmodified* test set. That is to say that although we are changing the threshold associated with the fault/no-fault classification in the decision tree, we are not changing the labels associated with the testing data, nor are we removing instances from the test set. The line on the plot shows that no modification can be made to the J48 threshold so as to obtain the same true positive/false positive rate as the 5-Stage Cascade.

To see if thresholding would permit either the One-vs-One SVM or the 3-Stage SVM Cascade to reach a similar performance profile as the 5-Stage SVM Cascade, we performed a similar analysis as for J48. Here, however, we applied the threshold to the fault instances in the training set, thereby removing the fraction of examples whose voltage deviations were more subtle than the threshold allowed. Again, this has the effect of removing measurements from PMUs geographically distant from the fault location from the training set. Once the new classifiers were trained, we then tested on the full testing set and plotted the resulting lines. The plot reveals that the One-vs-One SVM tends to outperform the 3-Stage SVM Cascade for false positive rates between roughly 0.8% and 4.5% respectively. However, like J48 and the 3-Stage classifier, it does not quite yield the same true-positive

TABLE III. CONFUSION MATRICES FOR: 1-VS-1 SVM (LEFT); 3-STAGE SVM CASCADE (MIDDLE); 5-STAGE SVM CASCADE (RIGHT)

True Classification \ Predicted Class	SLG	LL	3P	NF
SLG	7063	9	0	53
LL	192	207	0	75
3P	1	0	23	243
NF	469	50	0	10900

True Classification \ Predicted Class	SLG	LL	3P	NF
SLG	6361	54	2	708
LL	79	380	0	15
3P	1	0	259	7
NF	83	162	267	10907

True Classification \ Predicted Class	SLG	LL	3P	NF
SLG	6329	18	0	778
LL	89	252	0	133
3P	1	0	139	127
NF	83	53	60	11223

performance as the 5-Stage classifier. The 5-Stage Cascade overall performs best showing the lowest false positive rate with the full training set (upward facing triangular point). For false positive rates between $0.008 - 0.017$ the 5-Stage SVM Cascade and the 3-Stage Cascade perform relatively similarly. However, the 5-Stage Cascade once again dominates performance for lower false positive rates.

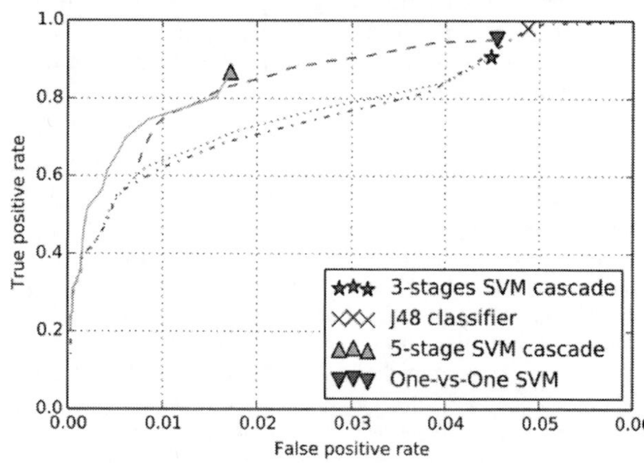

Figure 4. ROC curve for J48 and SVM classifiers

VII. LARGE TEST SET PERFORMANCE

The ROC curve in Figure 4 shows that no classifier outperforms the 5-stage SVM Cascade. In this Section we explore the 5-stage SVM Cascade's performance as we scale the test data to simulate a real operating environment. We expect that our previous testing is robust with respect to the system's ability to identify fault instances given that we have sampled an entire year's worth of line events that occur on transmission lines adjacent to one or more PMUs. However, our previous test set examined only a small sample of normal data. As a result, we expect that our estimate of false positive rate may be incorrect by a significant margin. Thus, in this experiment we focus on a contiguous 24 hours period in which no known line faults occur. Since line faults are relatively rare, this is not an atypical situation. Given this period of data, we then want to examine the performance of the 5-Stage SVM Cascade to determine how the false positive rate compares to our much smaller test set previously examined. To provide a baseline, we performed the same large scale test with the original J48 classifier.

We used signals from 19 PMUs across the grid over a contiguous 24 hour period. Each PMU reports 5,184,000 measurements over a 24 hour period, however one PMU is offline for 283 cycles (4.7 seconds) during the period we selected. Additionally, we use the first 10-cycles from each PMU to acquire a steady state value for each voltage measurement. In total this means that there are $(5.184 \cdot 10^6 \cdot 19) - 283 - (10 \cdot 19) \approx 98.5 \cdot 10^6$ individual examples to classify. However, from the analyst's standpoint, it is the moments of time that matter, not the individual measurements, so we consider only the unique moments where all PMUs report valid signals (i.e., $5.184 \cdot 10^6 - 293$ moments).

When run on these individual moments of time, the 5-Stage SVM Cascade classifier reports 15 false positive moments in time, but J48 reports an incredible 4,582,490 false positives. Our initial hypothesis was that one PMU may have been behaving erratically during the day we selected. To test this, we recorded the number of PMUs whose signals were classified as "fault condition" for each moment in time. We then applied a floating threshold so as to require multiple PMUs to corroborate each others' classification before marking a particular moment in time as a fault. We tested thresholds of 1,2,3, and > 3 PMUs. In the first case, we ask how many times exactly 1 PMU classifies a moment as a fault; in the last case, we ask how many times there are more than 3 PMUs classifying the same moment as a fault. These results are shown in Table IV. The first column shows how many moments cannot be classified (because they are used to initialize the steady state measurement or because a PMU is out of service); while the second column shows how many moments are classified as "normal" by all PMUs. The stratification demonstrates that J48 is not being adversely affected by a single PMU, since 75% of the day is classified as a fault by more than three PMUs.

Our next suspect was that J48 was adversely affected by the steady state measurement. Recall that one goal of our steady state computation was to ensure that we didn't inadvertently incorporate part of a fault into the window over which steady state value was calculated. As a result, our algorithm also ensured that moments in time labeled as a "line event" were not used in subsequent steady state calculations. This means that a contiguous string of false positives could result in the steady state measurement being significantly far back in time (and thus unlikely to be a precise measurement).

We re-ran the test using two variations of the steady state calculation, again recording the amount of corroboration between PMUs. The second steady state method used the same basic algorithm, but reported an error condition if the steady state window lagged more than three cycles behind the measurement point. This approach ensures that there was only minimal lag between the steady state window and the measurement, but may result in many errors (since there may be some periods that were not smooth enough to satisfy our steady state algorithm). Results from this approach are illustrated in Table V.

The third variation simply uses the previous 30 cycles for the steady state measurement, regardless of the smoothness of the measurements during that window, and regardless of the classification of data within that window. The attempt here was to ensure that the steady state window is always close in time to the measurement point and to reduce the time in which the steady state is incalculable. Since the window would always keep pace with the measurement, we expected that transient noise may create some false positives, but eventually the signal would smooth out and classification would continue regularly. We increased the window length from 10 to 30 cycles to avoid being overly influenced by isolated measurement outliers. Results from this approach are illustrated in Table VI.

TABLE IV. FALSE POSITIVES FOUND BY CORROBORATED CLASSIFICATIONS

	Out of Service	0 PMUs Report Fault	1 PMU Reports Fault	2 PMUs Report Fault	3 PMUs Report Fault	> 3 PMUs Report Fault
5 Stage SVM	293	5,183,692	10	3	0	2
Cascade	0.0057%	99.9941%	0.0002%	0.0001%	0.0000%	0.0000%
J48	293	601,217	139,196	226,854	285,764	3,930,676
	0.0057%	11.5976%	2.6851%	4.3760%	5.5124%	75.8232%

TABLE V. METHOD 2: LAGGED STEADY STATE WINDOW PRODUCES AN ERROR

	Out of Service	0 PMUs Report Fault	1 PMU Reports Fault	2 PMUs Report Fault	3 PMUs Report Fault	> 3 PMUs Report Fault
5 Stage SVM	333	5,183,653	10	2	0	2
Cascade	0.0064%	99.9933%	0.0002%	0.0000%	0.0000%	0.0000%
J48	1,763	5,181,655	527	26	11	18
	0.0340%	99.9548%	0.0102%	0.0005%	0.0002%	0.0003%

TABLE VI. METHOD 3: 30-CYCLE SS WINDOW KEEPS PACE WITH MEASUREMENT

	Out of Service	0 PMUs Report Fault	1 PMU Reports Fault	2 PMUs Report Fault	3 PMUs Report Fault	> 3 PMUs Report Fault
5 Stage SVM	313	5,183,669	13	3	0	2
Cascade	0.0060%	99.9936%	0.0003%	0.0001%	0.0000%	0.0000%
J48	313	5,181,968	1,440	130	41	108
	0.0060%	99.9608%	0.0278%	0.0025%	0.0008%	0.0021%

The results across all three steady state variations indicate that the 5-Stage SVM Cascade is relatively insensitive to the method of steady state calculation, while J48 performance is much more sensitive, even as the amount of corroboration required between PMUs is increased. Overall, the results from the 5-Stage Cascade are quite promising in this streaming environment. Across the 24 hour period, there are only 2 moments in time (regardless of the steady state method employed) that are incorrectly classified as faults by more than three PMUs. This false positive rate is 0.0000386% — more than four orders of magnitude less than expected given the results from the original test set.

VIII. CONCLUSIONS

In this paper, we have demonstrated a 5-Stage Cascade of SVMs within the PMU line fault domain. Our cascade offers a better performance profile than previously reported J48 classifiers for the same domain. Critically, we also have shown that the 5-Stage classifier achieves very low false positive rate on a 24 hour period of data containing almost 100 million examples. For future work, we are exploring the potentials of using unsupervised learning to characterize line events and unknown events on the smart grid.

ACKNOWLEDGMENTS

This work is partially supported by the DoE / the Bonneville Power Administration through the Technology Innovation Program (TIP #319). The authors would also like to thank BPA for the use of archival PMU data.

REFERENCES

[1] M. Hall, E. Frank, G. Holmes, B. Pfahringer, P. Reutemann, and I. H. Witten, "The WEKA data mining software: an update," ACM SIGKDD explorations newsletter, vol. 11, no. 1, 2009, pp. 10–18.

[2] R. Quinlan, C4.5: Programs for Machine Learning. San Mateo, CA: Morgan Kaufmann Publishers, 1993.

[3] Americans for a Clean Energy Grid, "Synchrophasors," http://cleanenergytransmission.org/wp-content/uploads/2014/08/Synchrophasors.pdf, 2014, [Retrieved: September, 2015].

[4] O. Antoine and J. C. Maun, "Inter-area oscillations: Identifying causes of poor damping using phasor measurement units," in Power and Energy Society General Meeting, 2012 IEEE. IEEE, 2012, pp. 1–6.

[5] K. Mei, S. M. Rovnyak, and C.-M. Ong, "Clustering-based dynamic event location using wide-area phasor measurements," IEEE Transactions on Power Systems, vol. 23, no. 2, 2008, pp. 673–679.

[6] O. P. Dahal, S. M. Brahma, and H. Cao, "Comprehensive clustering of disturbance events recorded by phasor measurement units," IEEE Transactions on Power Delivery, vol. 29, no. 3, 2014, pp. 1390–1397.

[7] R. Caruana and A. Niculescu-Mizil, "An empirical comparison of supervised learning algorithms," in Proceedings of the 23rd International Conference on Machine Learning. ACM, 2006, pp. 161–168.

[8] Y.-G. Zhang, Z.-P. Wang, J.-F. Zhang, and J. Ma, "Fault localization in electrical power systems: A pattern recognition approach," International Journal of Electrical Power & Energy Systems, vol. 33, no. 3, 2011, pp. 791–798.

[9] R. Diao, K. Sun, V. Vittal, R. O'Keefe, M. Richardson et al., "Decision tree-based online voltage security assessment using pmu measurements," IEEE Transactions on Power Systems, vol. 24, no. 2, May 2009, pp. 832–839.

[10] P. K. Ray, S. R. Mohanty, N. Kishor, and J. P. Catalão, "Optimal feature and decision tree-based classification of power quality disturbances in distributed generation systems," IEEE Transactions on Sustainable Energy, vol. 5, no. 1, 2014, pp. 200–208.

[11] R. Salat and S. Osowski, "Accurate fault location in the power transmission line using support vector machine approach," IEEE Transactions on Power Systems, vol. 19, no. 2, May 2004, pp. 979–986.

[12] F. R. Gomez, A. D. Rajapakse, U. D. Annakkage, and I. T. Fernando, "Support vector machine-based algorithm for post-fault transient stability status prediction using synchronized measurements," IEEE Transactions on Power Systems, vol. 26, no. 3, 2011, pp. 1474–1483.

[13] D. Nguyen, R. Barella, S. A. Wallace, X. Zhao, and X. Liang, "Smart grid line event classification using supervised learning over PMU data streams," in Proceedings of the 6th International Green and Sustainable Computing Conference (IGSC 2015), Dec 2015, pp. 1–8.

[14] C. Cortes and V. Vapnik, "Support-vector networks," Mach. Learn., vol. 20, no. 3, 1995, pp. 273–297.

[15] F. Pedregosa, G. Varoquaux, A. Gramfort, V. Michel, B. Thirion et al., "Scikit-learn: Machine learning in Python," Journal of Machine Learning Research, vol. 12, 2011, pp. 2825–2830.

Transforming Utilities: Turning Data into Intelligence with Data Analytics

Vanesa Čačković, Željko Popović
Ericsson Nikola Tesla did.
Zagreb, Croatia
e-mail: vanesa.cackovic@ericsson.com
e-mail: zeljko.popovic@ericsson.com

Abstract—**Data analytics is an important part in solving the challenges faced by transforming utilities. The introduction of smart meters and smart grids is leading to the convergence of information and communications technologies and utilities. Data analytics can be a powerful tool to create major improvements in efficiency and customer experience or it can even bolster service innovations. The deployment of data analytics is a necessary move as infrastructures, particularly electric grids, are expected to transform from more or less linear energy systems into networked, distributed systems with a multitude of market participants and management models in play. The grids simply become too complex to handle without large-scale data analytics.**

Keywords-data analytics; utilities transformation; smart metering; smart grid.

I. INTRODUCTION

Data analytics is a broad term frequently heard in the utility industry together with the phrase big data. In theory, analytics is simple: data delivered from devices, sensors, and meters throughout the grid is being analyzed and used to create an intelligence in order to inform and improve operator responses to a given situation. Most of the analysts agree that analytics solutions and big data have gone from hype to work [1][2][3][9][11]. It is becoming a foundation for decision-making and a business priority among the world's organizations. Big data is a term describing large amounts of data, collected from variety of sources, analyzed with the purpose of creating new intelligence and building business advantages. It brings faster results, a new depth to analytics and an unprecedented predictive power. Data analytics solutions are primarily used for three things: increasing customer experience, efficiency and growth. They are used as powerful preventive mechanism as well as corrective mechanism for real-time and near real-time control and management. A data-enhanced customer experience provides a deeper understanding of users and an increased customer experience through high performance services, fast feedback and customized offerings. Data-driven efficiency means taking advantage of the information available within the organization, in order to work smarter and reduce costs. Growth, finally, is about innovations and new revenue streams sparked by big data [4].

The utilities industry is transforming. Electrical grids, water and gas pipes are examples of infrastructures that are expensive to build and maintain, and are often heavily regulated and considered 'natural monopolies'. Market conditions vary significantly between countries in terms of competitiveness and private/public ownership, but in general, the sector has particular ways of conducting business. Simply put, utility businesses have been built around energy resources and have had generation, transmission and energy retail flowing from the center in a one-way, linear build-up. Consumers have had little influence on service and pricing. Today, there is a new situation for utilities companies: market mechanisms are changing, competition is increasing and pricing is becoming increasingly complex. There are a number of factors that contribute to the market transformation and call for big data solutions. Some of the most important drivers include:

1. Regulatory pressure: Governments are under pressure to improve energy efficiency, and their actions have consequences for energy companies. Regulations like the European Union's 2012 Energy Efficiency Directive and the US Energy Policy Act of 2005 put pressure on energy companies to become more efficient and reduce their environmental impact. Utilities are also incentivized to utilize their assets in more effective ways. This is done through value-based assets management, including deployment of sensors and analysis of the financial impact of outages.

2. Consumers become producers: The relationship between utility companies and consumers is becoming increasingly complex. The combination of decentralized generation and storage options with demand side flexibility can further enable consumers to become their own suppliers and managers for (a part of) their energy needs, becoming producers and consumers and reduce their energy bills. Decentralized renewable energy generation, whether used by consumers for their own use or supplied to the system, can usefully complement centralized generation sources. Where self-consumption exhibits a good match between production and load, it can help reducing grid losses and congestion, saving network costs in the long-term that would otherwise have to be paid by consumers. If consumers generate their own electricity from onsite renewable energy systems, they consume less electricity from the grid. This will affect how network tariffs are calculated. Network tariffs should be designed in a cost-reflective and fair manner while supporting energy efficiency and the renewable energy objectives and being simple and transparent for consumers Those dramatic changes in market mechanisms push energy

companies to find ways to manage distributed generation and net metering using the data analytics to help manage all new relations, power flow and business models.

3. Increase in price volatility: The increase of intermittent energy sources like wind and solar has resulted in more fluctuations in energy access over time. The volatility is elevated further with the growing influence of power markets. By introducing smart meters, consumers can adjust their energy use in relation to the fluctuating prices. These fluctuations are now becoming normal; but handling those demands require sophisticated data analysis.

4. Threats against the service delivery: Utilities are among the most critical infrastructures in society. Major incidents such as geopolitical events and natural disasters can cause severe damage to vital systems, in turn threatening markets, security, and people's lives. We all rely on energy infrastructure, and the increased level of these threats calls for improved resilience and intelligence.

5. Aging infrastructure: Around the globe, utility infrastructure is growing, and with aging technology comes increased vulnerability and a lack of interoperability with newer systems. Old infrastructure requires ongoing optimization to remain functional and cost effective [5].

Implementation of Distribution Grid Optimization Analytics described in this paper proposes using existing utility data fetched from several sources of smart metering network to achieve better visibility and easier monitoring of the end-to-end electricity network. It is important to state that analytics engine used here is real near time which enables users to act on the identified situation immediately. Another important aspect of this implementation is that the data model internally used by this analytics system is based on metadata which enables modelling of any kind of network elements, communicating over any communication technology in any type of network (electricity, water, gas networks with smart meters, routers, repeaters etc. communicating over power line communication, RF Mesh, 3G or any). So, the same concept could be applied for different types of utility networks.

This paper presents the typical current situation in European utilities regarding implementation of analytics methods based on industry research and commercial implementations experience together with possible benefits for implementing analytics and big data solutions.

The paper is structured in following way: Section I gives an introduction to the subject, Section II describes the Traditional utilities approach to data, Section III explains the Data Analytics Potential, Section IV gives an overview of the Implementation Use Case, while Section V gives the Conclusion of the paper.

II. TRADITIONAL UTILITIES APPROACH TO DATA

Today utilities only make use of existing data to a limited extent. Decisions are based mostly on historical data from internal data sources and sometimes combined with external source, for example weather data. Usage of real time data is limited to pilot projects and small scale initiatives. Utilities often do not use all valuable data they have access to. Major analysts estimate that around 5% of existing data is being used for some specific purpose, while the rest is not processed at all. For example, a smart meter can be seen as a sensor that can help to visualize "the health" of power grid and communication. Only small percent of data one smart meter is capable of storing and sending is being fetched and that is primarily for billing purposes. The fact that a utility collects data does not necessarily means that it is making use of the data. Everyday operation and maintenance routines often mean handling of various excel files coming from different data sources and is not being done in real-time manner. Databases are also often updated by hand.

Utilities approach can be described as "one problem/one system" approach which means that single problems are handled using single dedicated applications. Links between systems are loose or missing. This means that many utilities still lacks synchronized databases and use many redundant databases (data silos). However, efforts to integrate those separate data systems can be already seen for most of energy companies.

Today's utilities usually have some business intelligence tools in place and have insight into some areas of their operation. Those tools used by utilities are mostly standard reporting tools and obsolete forecasting models.

However, utilities are aware of the changes in the energy landscape and available data sources. They see that conversion of operation and information technologies brings massive volume of data. Some utilities have started pilot projects to investigate the use cases for big data other than accurate billing. It is clear that utilities will have to invest in new technologies for data management to unlock true potential of new data sources. They need to re-build their data architecture model and redefine data use cases.

Data management is not new for utilities. For example, traditional supervisory control and data acquisition (SCADA) systems [6] are in place for a while now, but the volume and velocity of data in smart grid is something which they are not prepared for. The main difficulty that they will have to overcome to start implementing big data technologies refers to expertise, infrastructure, and perception of risk (business value recognition).

Handling a tremendous amount of data from the grid, smart homes, in-house data bases or social networks requires big data processors with enormous storage capacity and specialized ICT services. Most utilities do not have that. They also complain about the shortage of people specialized both in energy industry and big data.

Business model transformation will mean step by step adoption of new data management technologies for more informed decision making. The transformation will be based on 3 pillars:

- Emergence and adoption of new data sources
- Integration of new types of data
- Changes in data consumption model

With the emergence of smart grids, utilities will start exploiting new data sources (grid sensors, smart meters, and electric vehicles) to optimize their business and provide better customer services. Utilities will be also forced to improve customer interaction building on social media and various web data.

Utilities are used to dealing with simple and structured data limited in volume and velocity. Collection, storage, and processing of M2M data from sensors and meters will pose a challenge for energy companies who have no in-home big data capabilities. However, the greatest difficulty will be to integrate those structured data with unstructured one which is totally new to utilities [1][7].

Business model transformation will be based on connecting traditional and new data sources, tapping new data, and integrating smart meter, smart grid, transaction, and geospatial data as shown on Figure 1.

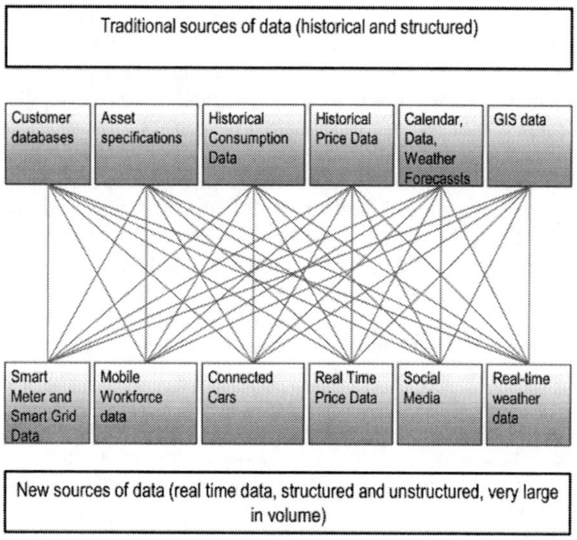

Figure 1. Data-business model transformation

Big data in the utilities sector will be dependent on a much larger number of nodes than for example in telecoms, demanding lots of machine power and powerful analytic models. Smart meters are not the only components of smart grids that can generate useful data for utilities companies.

In the transformation, period utilities will not be using the full potential of available data. Utilities will leverage data potential only to some extent, mainly using just smart meter data on commercial scale. Starting from reporting, billing and settlement, they will test new technologies and use cases. After choosing the most successful ones, they will start adding new sources of data to their data management systems and build up their analytics capabilities.

The development pace of data management systems will differ by country, energy market participants and technology.

Utilities will start with analyzing structured data. Many pilot projects, especially involving smart meter data have already been kicked off. Utilities will lag behind with adoption of analytical tools for processing unstructured data. It seems that cost approach will be the biggest challenge which will have to be overcome to enable moving to a truly data-driven model.

Cloud computing may be the answer for utilities hesitancy in deploying advanced analytics solutions. Some utilities have already moved part of their processes to cloud environment, however, a majority of those cases are non-mission critical processes.

III. DATA ANALYTICS POTENTIAL

There is evidence that analytics is quickly becoming an extremely important component of many utilities' overall IT/OT strategy. From predictive maintenance to customer engagement (CE), from energy efficiency optimization to grid optimization and beyond, analytics solutions are evolving rapidly and promise to dramatically alter the utility IT landscape in years to come. In a study commissioned by C3 Energy, the consulting group McKinsey and Company found that the potential annual savings per meter, through utility analytics solutions, could approach $300 [2].

By implementation purpose analytics solutions can roughly be divided into three categories:

1. Distribution Grid Optimization Analytics
2. Customer experience analytics
3. Business Intelligence analytics

1. Distribution Grid Optimization Analytics:

Smart grid data analytics can support a myriad of use cases, but in order to fully maximize the benefits of smart grid deployments, analytics solutions that support operational improvements are necessary. Grid optimization analytics can leverage data from smart meters and other grid-sensing devices, as well as through asset monitoring data, to optimize the operation of the transmission and distribution networks and asset performance [2].

For the Distribution Grid Optimization Analytics it makes most sense to include the analytics solution from the very start, with the rollout of smart meters. If so, the analytics solution can be powerful tool to track the rollout progress. Main benefits of using analytics in parallel with the rollout are reducing times for detecting and fixing faults, decreasing number of needed field interventions, increasing quality of service, improving network visualization and decreasing the operational cost of the network in general. After the rollout is finished, with the analytics solution already in place, the same system can be used for further distribution grid optimizations (for example, Demand Response Management System (DRMS), energy data forecasting/load forecasting, various visualization tools etc.).

2. Customer experience analytics:

The business-to-consumer model is being challenged by forms of energy production that allow private individuals to become energy producers. Wind farming, solar energy, the unloading of surplus energy from electric vehicles into the grid, or other forms of distributed generation (the ability for consumers to sell power) all complicate the transaction of energy and payments. This in turn demands more from billing and charging systems. Data analytics is necessary for dealing with the complex customer relationships following from distributed generation. It can also provide the foundation for new services that address new customer needs. Smart meters and smart home appliances provide new opportunities to invent solutions that build customer engagement and grow revenue. They can track consumer behavior on different appliances, and combine that information with weather data, pricing information and other variables to create solutions that save energy and money for users. Data analysis can be used to mitigate price volatility on the consumer level, through better management of the risks associated with a portfolio of commercial and physical assets. These types of data-driven innovations help to build new services, improve billing and charging algorithms and ultimately build engagement among customers in the long term. They can help to transform utility companies' businesses beyond distributing and selling electricity.

Like telecoms, the utilities industry is a sector where timing is vital. The intricate network of assets needs continuous monitoring and support, especially as many infrastructures are aging, and ongoing optimization becomes a way of prolonging their life. Data analytics solutions can give companies early warning services, detecting signs of problems that may affect customer experience in advance. The aim is to optimize everyday performance and prevent issues through effective fault management. There is also a customer care dimension to experience issues. Through analytics solutions, companies can identify the underlying issues that lead to customer complaints. It becomes possible to drill down into each customer's usage history and find the root cause of issues, making it possible to explain and resolve bad experiences.

3. Business Intelligence analytics:

Broadly, business analytics are used to guide business planning and identify inefficient business processes. Within the utilities sector, business analytics have historically been used for financial and production planning applications to analyze historical data and provide information to business users around a set of predefined key performance indicators via reports, dashboards, and web portals. Typical applications include financial and compliance reporting, demand forecasting for energy trading, and customer operations. Today, armed with a multitude of new data generated by smart grid technology deployments, utilities are trying to understand the potential for business intelligence (BI) data analytics applications beyond those provided by traditional BI tools. BI analytics can leverage smart meter data and asset monitoring data to improve the efficiency of M2C operations, enhance revenue assurance, provide greater visibility into system performance and asset utilization, and enable load and generation forecasting for energy trading or production planning activities. The more advanced applications move from a predominantly descriptive approach toward more predictive analytics that will enable utilities to become proactive in their strategies [2].

IV. Implementation Use Case

Implementation of analytics solution as a rollout support is being chosen as an example for this paper since it describes well the usefulness of the analytics solution for utility company as well as help to understand basics and the full potential on one analytics solution.

A. Architecture

Advanced Metering Infrastructure (AMI) is defined as a full measurement and collection system that consist of a meter hardware located at the customer site; communication networks between the customer and a service provider such as an electric, gas, or water utility; and data reception and management systems that make the information available to the service provider. AMI has a variety of specific characteristics that present challenges to the effective management of its communications and head-end systems. The biggest challenges an AMI network faces include controlling and utilizing waves of meter data, building AMI networks to scale, and meeting important requirements including bandwidth, latency, throughput and reliability.

Advanced Metering Infrastructure are systems that measure, collect, and analyze energy usage, and communicate with metering devices such as electricity meters, gas meters, heat meters, and water meters, either on request or on a schedule. These systems include hardware, software, communications, consumer energy displays and controllers, customer associated systems, Meter Data Management System (MDM) and supplier business systems.

The most common architecture used to address the features of smart metering systems is presented in Figure 2. It consists of [11]:

- Smart meters (SM)
- Data concentrators (DC): process data from several meters
- Head End System (HES): central data collection point
- Local area network (HAN, NAN): allows bi-directional communication between the smart meters and a data concentrator
- Wide area network (WAN): allows bi-directional communication between the data concentrators and the head end system.

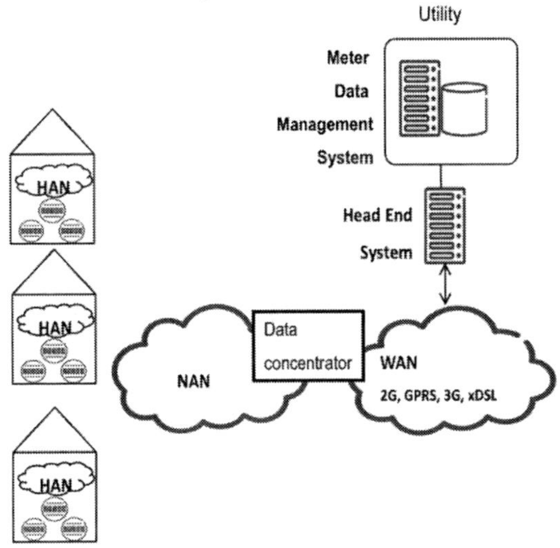

Figure 2. Typical AMI architecture

Though the basic function of the MDMS will include validation and storage of smart meter data, utilities expect additional functionality in support of more business processes across their value chain. Meter Data Analytics (MDA) is an upcoming trend where Meter Data Management (MDM) vendors or ICT vendors are trying to expand their offerings into in order to provide utilities add-on capabilities for data processing, analyzing, and reporting trends that add value rather than just uploading data. MDA enables utilities to avoid forecast failures rather than respond to them.

B. Implementation

Utility company described in this use case is one of the largest electricity distribution companies on the Baltic countries, owning over 640.000 metering points (residential and industrial), 61.000 km of power lines and 22,000 substations. Main motivation for transformation the utility was the fact that all electricity meters need to be remotely readable by 2020 according to EU directive and aging equipment with high maintenance cost. The utility company started an eight-year long transformation project within which integration of new MDMS was included as well as rollout of smart meters and integration of Analytics platform with existing systems. Smart meters help modernize the power grid by providing the ability to remotely monitor and transfer energy consumption and power quality information.

Smart metering also benefits end customers in the form of energy and cost savings, reduced carbon emissions, improved service reliability and improved customer service and responsiveness. To make all of these benefits a reality, utilities need a secure and flexible two-way communications infrastructure to connect and communicate with smart meters.

Collection and transmission of energy consumption data is a continuous process that is done automatically. As the network grows, efficient operation of Smart meter communication network becomes a critical process. Ability to address critical issues in network and also capability to address problems before a failure require innovative tools to support communication personal in their daily work. Analytics platform is a solution for enabling visualization and context development for smart metering data to support power grid management [12]-[17].

The Analytics Platform gathers the data from various data sources (HES, MDMS, GIS, CIS, asset management, etc.) in online mode and does the modelling and processing of data in order to act on newly processed information. During the implementation over 50 candidates for analysis and implementation were identified (so called use cases) which can roughly be divided into three groups: power network operations, smart meter operations and fraud&loss related use cases.

Examples of the use cases are:

1. Use Case (Communication supervision):

Loss of meter data during the collection process may cause issues with billing, as well as disturbances in the distributed energy generation. Upon a data collection failure, it is possible to quickly analyze possible causes by drilling down in the communications network, from a geographical network map down to an individual meter in an apartment house. Regardless where the information come from (multiple systems), it is presented in the same user interface which uses latest usability principles to increase user efficiency and effectiveness. The majority of the faults can be identified and solved either with remote reconfiguration or by deploying field personnel by issuing a work order. Network planning issues are found and can be corrected during normal operation. The method can be used to discover causes for income drops, users complaints, users churn, loss of control of distributed generation figures etc. There might even be regulations which put requirement on fault identification and fault repair which forces the utility company to have an efficient tool to localize and correct faults within a predefined period implying that every saved second means saved money.

2. Use Case (Power Quality Visualization):

Power quality determines the fitness of electric power to consumer devices. Without the proper power, an electrical device (or load) may malfunction, fail prematurely or not operate at all. The complexity of the system to move electric energy from the point of production to point of consumption combined with weather variation, generation, demand and other factors provide many opportunities for the quality of supply to be compromised. Smart meters provide simple power quality measures. Analysis of this information can aid in the detection of anomalies and other emerging adverse conditions in the distribution network, allowing the utility to proactively initiate equipment maintenance, reducing the likelihood of an outage. Power quality monitoring helps in preventive system repairs/ maintenance by identifying areas of poor quality and subsequently acting on those, additionally life of assets could be extended and unplanned service interruptions could be avoided via predictive and pro-active maintenance

3. Use Case (Energy Loss Detection):

Fraud amongst the subscribers is always a risk, including loss of income to unbalance in the power grid. Aggregation and correlation of multiple metering points (e.g. substation vs. aggregated feeder line consumption) makes it possible for the utility to understand magnitude of, and identify possible fraud activity. All information available from different sources is gathered in the same user interface to optimize the user experience and increase the efficiency. An alarm as well as a work order can be issued upon detected (suspected) fraud. The method can be used to reduce the loss of income but also prohibit unwanted unbalance in the power grid.

4. Use Case (Outage management visualization):

Smart meter events such as 'last gasp' and 'power restore' that provide meter off/on status can be used for improving outage management. Being real time or near real time events, these have an obvious advantage over outage information coming from customers and field staff. Examples of scenarios which could be visualized: Momentary outages and restoration-related events; communication and network interface issue-related events; Events due to planned outages, outages at the lateral, feeder or transformer level, customer disconnects, etc. Effective events processing & analytics engine shall be used to detect outages at the right device level, create proactive tickets, as well as 'power restore' to identify nested outages after large-scale outage restoration. Aim is to help make outages shorter and restoration more reliable. It can also reduce the customer complaints if information about the outage can be handled and in some cases distributed, even before the customers notices the outage.

After integrating described solution utility doubled the back-office engineer efficiency, times for solving faults were cut by half and overall communication quality between SM network elements improved significantly [8].

It has been seen that it would be beneficial to for use this type of Analytics platform also in other parts of utility organizations such as call center and general performance monitoring of the company [18][19].

As an outcome of the integrating the Analytics platform with existing systems and various data streams, it is possible to visualize the network, get communication quality, drill down to root cause, read configuration of devices and have near real time data available. The platform enables proactive and/or fast actions in order to secure needed network quality and meter data availability, supervise network rollout with higher quality, handling big number or smart meters and reduce operating costs. Analytics platform provides a way to collect, aggregate and analyze data in real-time, allowing utilities companies to prevent issues and manage the customer experience.

V. Conclusion

Collection and transmission of measurements and events from devices is a continuous process that is done automatically. The variety, velocity and volume of data gathered from and about utility networks are growing day by day. Such trend enables, but at the same time makes it difficult to process the data in order to make accurate, adequate and actionable decisions on managing those networks. Understanding the collected data brings opportunities to acquire new markets, customers and work with efficiency. Many organizations have a lot of data available, but do not have enough experts to leverage big data, and are not recruiting enough talent either. What is equally serious is that many organizations do not have good systems, routines and do not ascribe sufficient importance to data to make insights available to the whole organization.

Data analytics makes it possible to collect real-time data from existing grid elements to gain better control of the infrastructural assets. Ongoing monitoring of resources improves proactive maintenance actions and operations. Data from smart meters and other smart assets helps us to understand asset usage and provides a means of asset lifecycle management. The ability to anticipate issues that could interrupt the provision of service is a value that big data brings to both customer experience and operations. Like many other aspects of running a utilities operation, fault management is an area that is more difficult to handle today without data analytics solutions. Predictive modeling of events that cause outages can be deployed to trigger automated mitigating actions. This is an effective tool to reduce costs and improve customer experience.

References

[1] Smart metering in Europe; Tobias Ryberg, Berg Insight research team; www.berginsight.com, [Last access April 2016]

[2] Navigant research: "Smart Grid IT Systems MDMS, CIS, GIS, SCADA, EMS, DMS, OMS, AMS, MWMS, DRMS, DERMS, and Utility Data Analytics Solutions: Global Market Analysis and Forecasts"; published 2Q 2015

[3] Frost&Sullivan: "Data-driven Utilities: Business Intelligence and Analytics to become the Top Investment Area for Energy Companies", October 2013

[4] Industry Transformation - Horizon Scan: ICT & the Future of Utilities; http://www.ericsson.com/news/141211-horizon-scan-ict-and-the-future-of-utilities; [Last access April 2016]

[5] Actionable intelligence Transforming Utilities; Ericsson; http://www.ericsson.com/res/docs/2015/actionable-intelligence-transforming-utilities.pdf; [Last access April 2016]

[6] http://whatis.techtarget.com/definition/SCADA-supervisory-control-and-data-acquisition; [Last access April 2016]

[7] Data-driven Utilities: Business Intelligence and Analytics to become the Top Investment Area for Energy Companies; Frost&Sullivan; October 2013

[8] http://www.ericsson.com/industry-transformation/energizing-estonia/; [Last access April 2016]

[9] Rob Lockhart: "Smart Grid Data Analytics: a New Approach for Utilities"; Gigaom Research, 2015

[10] http://www.metering.com/smart-meter-europe-estonia-utility-deploys-half-of-630k-target/; [Last access April 2016]

[11] GTM Research: "The soft grid 2013-2020: Big Data & Utility Analytics for Smart Grid";

[12] "Functional reference architecture for communications in smart metering systems", CEN/CLC/ETSI/TR, Tech. Rep., ICS 33200; 91.140.01, Ref. CEN/CLC/ETSI/TR 50572:2011 E, December 2011.

[13] J. Kwac, J. Flora and R. Rajagopal, "Household Energy Consumption Segmentation Using Hourly Data", IEEE Transactions on Smart Grid, vol. 5, no. 1, January 2014, pp. 420-430.

[14] G. Chicco, "Overview and Performance Assessment of the Clustering Methods for Electrical Load Pattern Grouping", Energy, vol. 42, no. 1, June 2012, pp. 68-80.

[15] L. Hernandez, et al. "A Survey on Electric Power Demand Forecasting: Future Trends in Smart Grids, Microgrids and Smart Buildings", IEEE Communications Surveys & Tutorials, vol. 16, no. 3, Third Quarter 2014, pp. 1460-1495.

[16] I. S.Bayram, G. Michailidis, M. Devetsikiotis, and F. Granelli,"Electric Power Allocation in a Network of Fast Charging Stations", IEEE Journal on Selected Areas in Communications, vol. 31, no. 7, July 2013, pp. 1235-1246.

[17] A.G. Phadke and J.S. Thorp, "Synchronized Pharos Measurements and Their Applications (Power Electronics and Power Systems)", Springer, 2008, ISBN 978-0-3877-6535-8.

[18] Y. Zhang et al. "Wide-Area Frequency Monitoring Network (FNET) Architecture and Applications," IEEE Transactions on Smart Grid, vol. 1, no. 2, September 2010, pp. 159-167.

Multi-objective Optimization of Energy Hubs at the Crossroad of Three Energy Distribution Networks

Diego Arnone, Massimo Bertoncini,
Giuseppe Paternò, Alessandro Rossi
Research and Development Laboratory
Engineering Ingegneria Informatica S.p.A.
Palermo, Italy
e-mail: diego.arnone@eng.it, massimo.bertoncini@eng.it,
giuseppe.paterno@eng.it, alessandro.rossi@eng.it

Mariano Giuseppe Ippolito,
Eleonora Riva Sanseverino
Dipartimento di Energia, Ingegneria dell'Informazione e
Modelli Matematici (DEIM)
Università degli Studi di Palermo
Palermo, Italy
e-mail: marianogiuseppe.ippolito@unipa.it,
eleonora.rivasanseverino@unipa.it

Abstract—**This paper provides a multi-objective optimization framework aimed at the management of a multi-carrier energy system involving both electricity and hydrogen. Using the concept of the multi-carrier hub, the proposed system has been modelled in order to define completely every energy flow inside the plant. After that, a heuristic multi-objective optimization algorithm, the Non-dominated Sorting Genetic Algorithm II, has been implemented for the energy management of the plant, taking into account simultaneously three different objective functions related to economic and technical goals. This optimization process provides the set point defining the working configuration of the plant for a daylong time horizon. The communication framework between the energy management system, the real plant and the monitoring tool has been developed too, using the Open Platform Communications (OPC) protocol for the data exchange. This has been presented along with the Decision Support System (DSS) provided by the optimizer and the Human Machine Interface (HMI) of the Supervisory Control And Data Acquisition (SCADA) monitoring the plant. All the presented applications are going to be deployed on a real plant demonstrator.**

Keywords-multi-carrier hub; multi-objective optimization; energy management system; OPC protocol; hydrogen storage.

I. INTRODUCTION

Today, multi-carrier energy hubs are a concrete reality for the energy distribution networks management, enabling the interconnection between these energy infrastructures by means of several energy devices able to convert, store and buffer various forms of energy. In the view of remodeling and restructuring energy infrastructures, especially electrical ones, multi-carrier hubs could represent an innovative and cutting-edge technology for the design and realization of a new hybrid, flexible and interoperable distribution grid framework.

The concept of multi-carrier hubs was introduced by G. Andersson and et al. [1][2], defined as "units where multiple energy carriers can be converted, conditioned and stored; such a system represents an interface between different energy infrastructures and/or loads". They faced the problem from many points of view, proposing a graphic model and a complete matrix model able to correctly represent every device in the system and its operation. This research laid the foundation for the systematic study of multi-carrier hub framework in several energy grid domains. In the last two years, many researchers have addressed the challenge of multi-carrier hub. Besides the modelling criteria, which are mostly based on G. Andersson studies, many authors faced the issues of the control and management of a multi-carrier hub. These tasks are very frequently addressed by means of heuristic optimization processes, e.g., multi-objective optimization [3], fuzzy logic systems [4], multi agent systems [5] or an effective combination of two of them [6]. Another significant topic related to the operational issues of a multi-carrier hub is the management of the dynamic behaviour of two or more energy carriers that must be controlled as a whole inside the multi-carrier structure. Most of these studies focused on the interaction between electricity and thermal energy, because of the great difference between the time scales of these energy carriers [7][8]. Finally, some studies proposed a widespread usage of the multi-carrier hub concept over a city/district level in order to create a unique interoperable energy distribution network by connecting different energy infrastructures through two or more multi-carrier hubs properly located [9][10].

The present paper proposes an Energy Management System (EMS), based on a multi-objective optimization process, which is able to perform the management, scheduling, control and monitoring tasks of a multi-carrier system. This system involves two energy carriers, electricity and hydrogen, and interconnects a Medium Voltage (MV) electrical distribution grid, a Low Voltage (LV) electrical distribution grid and the methane/natural gas network for hydrogen delivery. It also integrates Renewable Energy Sources (RES) in the management of electrical power flows. The structure of such a system roughly consists of a Water Electrolyser (WE), connected to the MV grid and able to produce hydrogen, an innovative Hydrogen Solid-state Storage System (HSS), installed on two different channels, and a Fuel Cell (FC), connected to the LV grid, that absorbs hydrogen from one of the HSSs. The research performed on this system is part of the European co-funded INGRID project (7th Framework Programme) [11][12]. In Figure 1, the block diagram of an INGRID system instantiation is shown.

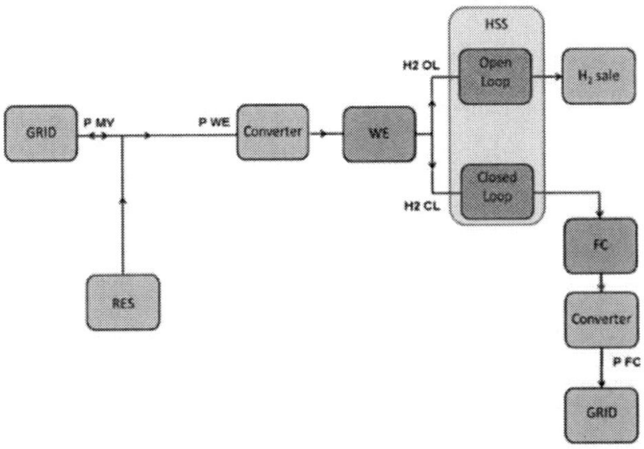

Figure 1. INGRID plant block diagram detailed with electricity and hydrogen flows.

The present paper is structured as following. In Section II, the INGRID system structure and operation are described and analyzed. In Section III, the energy management framework and its optimization process are depicted, along with the Objective Functions (OFs) and constraints. Section IV shows the communication framework and the EMS interaction with the monitoring tools, whilst Section V shows the optimization results and the EMS Graphic User Interface (GUI). In Section VI, the conclusions are given.

II. INGRID Project System

A. INGRID project

The INGRID project aims at contributing to balance supply and demand of different energy carriers. The main goal of this project is to handle the very large amount of power generated by RES systems, installed on MV distribution grid, by absorbing electric power, which is used to produce and store hydrogen [11]. In this way, an INGRID plant can prevent grid technical issues and power reverse flow phenomena. The electric energy absorbed by the WE is converted in hydrogen that is stored into two different storage systems belonging to two different channels: the Open Loop (OL) channel, in which the hydrogen is stored and then sent to the methane pipeline network, and the Closed Loop (CL) channel, in which the stored hydrogen is employed to supply the FC. The FC, in its turn, generates electric power for LV balancing services.

The EMS here proposed is tailored on the INGRID project demonstrator, which is being deployed, set up and will operate in Troia (Puglia, Italy).

B. INGRID EMS overview

The EMS of the INGRID system consists of several different components.

The core of the EMS is the Energy Supply and Demand Matcher (ESDM), which is in charge of scheduling tasks over 24 hours, performed by means of the optimization process. The flexibility of the optimization allows the user of the plant to properly choose even a shorter time horizon. It provides the power absorption and generation profiles for all the devices inside the system.

The data related to the forecasted power profiles of RES production and the prevision of price profiles are evaluated by a simulation tool that provides the EMS with these initialization data. Moreover, in order to meet the balancing strategy of the Distribution Service Operator (DSO), the grid operator sends to the EMS a suggested power consumption profile for MV grid and a suggested power generation profile for LV grid before the optimization process.

The EMS is also responsible of the monitoring of the entire plant by means of a specific tool developed in WinCC® [13] environment. The monitoring tool consists of an integrated HMI, that collects all the equipment operational parameters, as well as the alarms, and a communication framework which adopts the OPC protocol [14]. In Section IV, the monitoring tool is depicted.

At the end of the optimization process, a DSS allows the human operator to interact with the EMS and choose the most suitable working configuration for the plant. The DSS therefore selects a fixed number of optimized solutions and provides the user with a Graphical User Interface (GUI) through which these solutions can be displayed, compared and then adopted. These tools are shown in Section V.

III. Multi-objective Optimization Framework

A. Non-dominated Sorting Genetic Algorithm II (NSGA-II)

The optimization process is implemented by means of an evolutionary algorithm, the Non-nominated Sorted Genetic Algorithm II (NSGA-II) [15][16]. Such an algorithm allows to simultaneously optimize two or more objective functions simulating the biological phenomenon of the evolution. The NSGA-II offers good performance in terms of convergence, as well as scalability, as it is shown by adding a new third objective function to the original optimization framework. The total complexity of the algorithm is $O(mN^2)$, where m is number of the objective functions and N is the size of the considered population.

This algorithm has been already tested and acknowledged in previous studies on the field of smart grid and electrical network management [17][18][19]. Moreover, its efficacy has been evaluated by comparing it with previous studies [20] that have addressed INGRID plant optimization task by means of mono-objective heuristic optimization processes, such as Tabu Search and Simulated Annealing.

Such a kind of optimization process offers a set of solutions, which is the population front resulting from the last generation. The DSS will help the user of the INGRID EMS to properly choose the most suitable solution.

B. EMS optimization

The optimization carried out by the ESDM module aims at scheduling all the electric power and hydrogen flow profiles inside the plant while optimizing two or more objective functions which are strictly related to the operational conditions of the system equipment.

The three INGRID system parameters selected for being optimized by the NSGA-II are the hydrogen flow in OL channel, the hydrogen flow in CL channel and the electric power generated by the FC. It worth noting that the sum of OL and CL hydrogen flows defines the power consumption of the WE through the efficiency of this device. In this way, all the significant parameters of the plant are managed, since OL hydrogen flow is directly linked to the State of Charge (SoC) of the HSS installed in this channel and thus to the hydrogen produced to be sold. Similarly, the CL hydrogen flow is responsible of the SoC of the HSS in the CL channel, which in its turn provides the generation availability of the FC. The complete definition of these physical quantities, achieved by implementing the multi-carrier hub model [21], allows to set three objective functions strictly related to them.

The first objective function is based on economic criteria. It addresses the maximization of the daily revenues deriving from the sale of hydrogen to the hydrogen market and balancing services to the LV grid. It can be defined by roughly considering the purchase cost of the input energy carriers, i.e. electricity from MV grid, and a sale price of the output energy carriers, i.e. electricity to LV grid and hydrogen to its proper market:

$$OF1=\sum_{i=1}^{24} - \left\{ \begin{array}{l} \left[\left(P_{e,WE}(t_i)-P_{e,RES}(t_i)\right)c_{grid}(t_i)\right]+ \\ -\left[L_{H2}(t_i)p_{H2}+L_{e,LV}(t_i)p_{ANC}(t_i)\right] \end{array} \right\} \quad (1)$$

where:

- $P_{e,WE}$ is the electric energy consumption of the WE during a time step, [kWh];
- $P_{e,RES}$ is the electric energy generated by the internal RES system during a time step, [kWh];
- L_{H2} is the amount of produced hydrogen to be sent to the H2 market during a time step, [kg];
- $L_{e,LV}$ is the electric energy produced by the FC injected into the LV grid during a time step, [kWh];
- c_{grid} is the energy purchase price from MV grid, [€/kWh];
- p_{H2} is the hydrogen sale price, [€/kg];
- p_{ANC} is the electrical energy sale price to the LV grid for balancing services, [€/kWh].

The second objective function is instead related to the smart grid philosophy adopted by the INGRID system: one of the most significant goal is to support DSOs on coping with power flows imbalances mainly caused by the huge amount of power produced by RES installed on MV distribution grid. In order to avoid RES generation curtailments, a very expensive and deplorable practice, the DSO estimates a power consumption profile that fits its technical contingencies and should be followed by the INGRID system. The EMS is therefore asked to accomplish this goal without constraining system operation to a fixed power value: actually, the WE is not forced to absorb the electric power suggested by the DSO, but its power consumption depends on the optimization strategy, that takes into account DSO. This has been implemented by means of a technical objective function aiming at minimizing the distance between the DSO power profile and the real power absorption of the INGRID system:

$$OF2=\sum_{i=1}^{24}\left[\frac{\left(P_{grid,MV}(t_i)-P_{DSO}(t_i)\right)}{1000}\right]^2 \quad (2)$$

where:

- $P_{grid,MV}$ is the total power absorbed from the MV grid, [kW];
- P_{DSO} is the power consumption suggested by the DSO, [kW].

This objective function as been designed as a numeric quadratic index in order to address suitably the distance between the two power profiles.

The evaluation of daily economic revenues takes also into account the level of compliance of the INGRID plant with this curve: if the plant manages to satisfy DSO within a small range around the suggested profile, a price discount for energy purchase is considered.

In this paper, a third objective function is introduced in order to manage the power injection of the FC in the LV distribution grid. As seen above, the request of a power profile, either a generation or a production one, is not considered as a constraint but as a suggested behaviour to fulfil network operators strategy. LV power profile is evaluated to exploit the availability of the INGRID CL storage system for, e.g., balancing services and for electric vehicle recharge programs. This third objective function is shaped as the second one, being a numerical index that stands for the distance between the power generation request and real power produced by the FC:

$$OF3=\sum_{i=1}^{24}\left[\frac{\left(P_{FC}(t_i)-P_{LV,dem}(t_i)\right)}{1000}\right]^2 \quad (3)$$

where:

- P_{FC} is the power produced by the FC available for LV balancing services, [kW];
- $P_{LV,dem}$ is the power demanded by DSO, [kW].

For the sake of simplicity, the constraints of the optimization problem are just described. These constraints are mainly inequality ones. They are related to the maximum and minimum rated power of the WE and FC, the maximum and minimum hydrogen flow of the OL and CL channels, the capacity of the two HSSs. Other technical constraints are due to the power variation limits imposed by the WE and FC. Actually, the delta power between two time stamps cannot be larger than a fixed value. Finally, the new hydride technology of HSSs needs particular conditions and procedures for the absorption and desorption tasks, which are taken into account by means of operational constraints.

As already stated, the requests of power profile at the MV and LV grid interface are not handled as equality constraints but by means of two different objective functions, so these profiles are followed according to the optimization criteria. This is one of the most innovative concept proposed by the INGRID project.

IV. Communication Framework And Monitoring Tool Integration

In this section, the communication framework among the EMS, the monitoring tool and the real plant devices are outlined. In Figure 2, all the modules of this framework and their interconnections are shown.

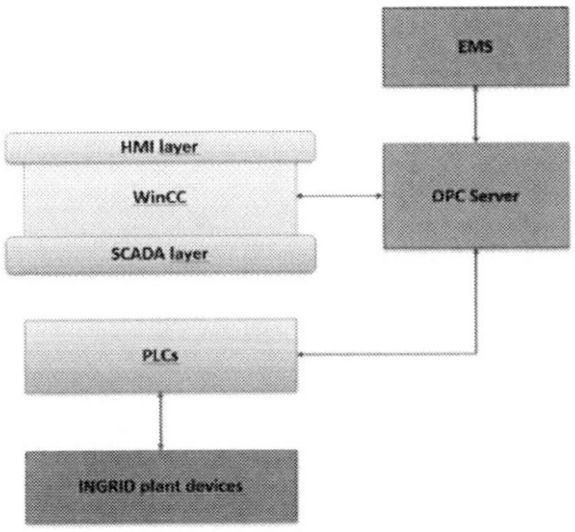

Figure 2. INGRID communication framework.

The OPC server has a central role in this structure since all the data are exchanged through the OPC protocol. In order to start the optimization process for the desired time horizon, the EMS reads the initialization data and the current plant configuration from the server. The data pertaining the device are synchronized to the Programmable Logic Controller (PLC) of each device inside the plant, and is also employed for the monitoring tasks handled by WinCC®. In its turn, the EMS performs the optimization and provides the system with the set points for the each time stamp over the entire time horizon. These set points, chosen from a continuous domain, are written in the OPC server and then are used by the monitoring tools, as well as from the PLCs of the real plant as an input for the real devices.

The optimization framework is able to cope with deviations of the real plant configuration from the set points suggested by the EMS or with errors in forecasted data. In this case, the optimizer performs a new optimization in real-time, starting from the data of the current configuration and/or the new forecasted profiles; the time horizon can be set considering only the remaining hours of the day.

The monitoring tool is a component developed on purpose for the INGRID project, which collects and integrates all the set points, operational parameters and the alarms regarding the plant equipment. In Figure 3, its HMI is shown. The HMI is part of the SCADA of the plant, which can be used for the general control of the plant. The EMS, by means of the DSS, can access the SCADA to perform its optimization. Human intervention is always prioritised, as it can be expected.

The HMI is currently implemented as a demo, since the OPC server is interfaced by simulation drivers.

V. RESULTS AND GRAPHIC USER INTERFACE

As mentioned in the previous sections, the optimization process schedules the set points of the plant equipment for a user defined time horizon. These set points are selected by the algorithm in order to achieve the best values of the OFs, thus the suggested plant configuration should allow high economic revenues, a good MV profile following and a good LV profile following. Nevertheless, the algorithm provides an optimized solution front, made by a number of suitable solutions equal to an entire population. Each of these solutions implies different values of the OFs, so it is important to properly choose the best solution that fits the user's current goal.

A DSS has been realized in order to ease this task: it automatically selects the three solutions that allow to reach the best value of the first OF (OF1 best), the best value of the second OF (OF2 best) and the best value of the third OF (OF3 best). They are clearly outlined by the GUI of the optimizer application, shown in Figure 4. This GUI displays some initialization data and the starting plant configuration, all the optimized energy flow profiles inside the plant for the three solutions selected by the DSS, the price profiles and the forecasted profiles of RES. The real plant data are not available yet, since the demonstrator plant is still under construction; so, the initialization profiles and the starting plant configuration are supposed among those representing a possible and effective operating condition.

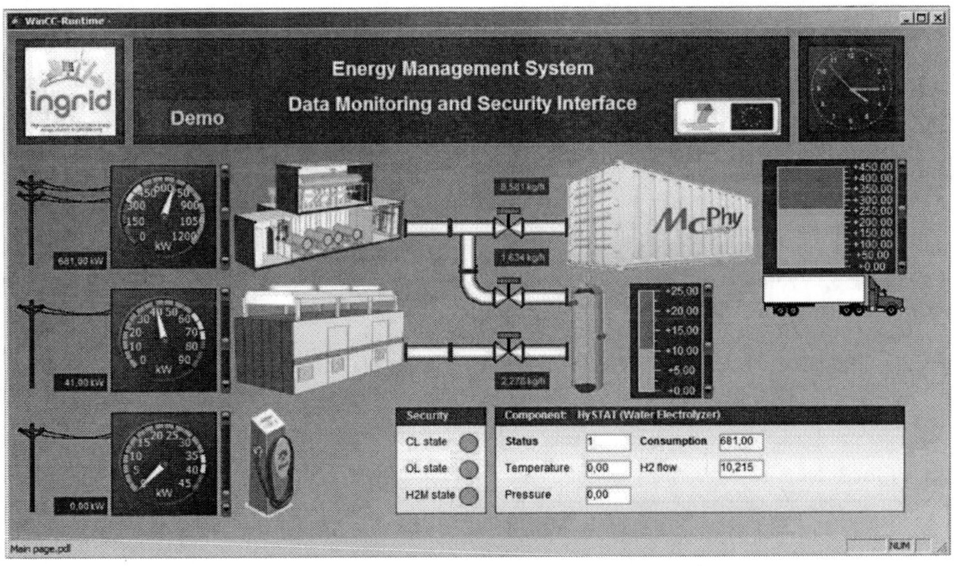

Figure 3. The HMI of the SCADA of the plant developed in WinCC® environment.

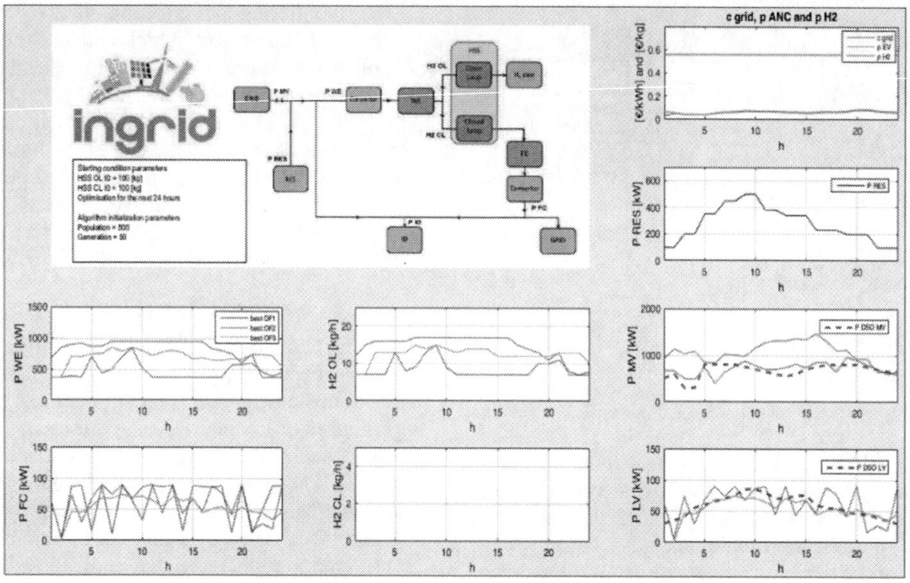

Figure 4. The Graphic User Interface (GUI) of the optimizer application developed in MATLAB® environment.

In particular, the last two rows of the GUI show the power consumption of the WE (P WE), the hydrogen flow on OL channel (H2 OL), the hydrogen flow on CL channel (H2 CL) and the power produced by the FC (P FC). All these data are provided for the OF1 best solution (blue lines), OF2 best solution (red lines) and OF3 best solution (green lines).

The last two charts of the last column show the graphic comparison between the two power profiles requested by the DSO, both for MV and LV grid, as well as the actual power absorption by the INGRID plant from MV grid and the power injected on LV grid. For the sake of clarity, they are shown again in Figure 5. It is possible to notice how INGRID plant manages to approximately follow the requests of the DSO. It worth noting that the best adherence to the MV profile request is achieved by means of the OF2 best solution, whilst the best adherence to the LV profile is achieved by means of the OF3 best solution. In these two graphs, the OF2 best curve (red) and the OF3 best curve (green) are compared to the OF1 best curve (blue), in order to show how OF1 best solution does not allow to respond to DSO profile suggestions (purple).

The human operator of the plant, once examined these solutions, has to choose the one that fits better the current operational contingencies of the plant. Depending from economic advices or grid technical issues, the human operator can be oriented on adopting a solution instead of another one. On this purpose, the DSS provides the user with a GUI that allows to choose one of the proposed solutions, making it the current desired plant configuration. In Figure 6, this GUI is shown. Using this interface, the user can accurately analyze the power profiles of the WE and FC, selecting them from the pop-up menu, for the three solutions selected by the DSS. The numerical values of the three OFs are displayed, too. Once selected the desired solution, the user is asked to click on the "OK" button to make it effective and to send all the configuration data to the OPC server to be synchronized to the PLCs controlling the plant devices.

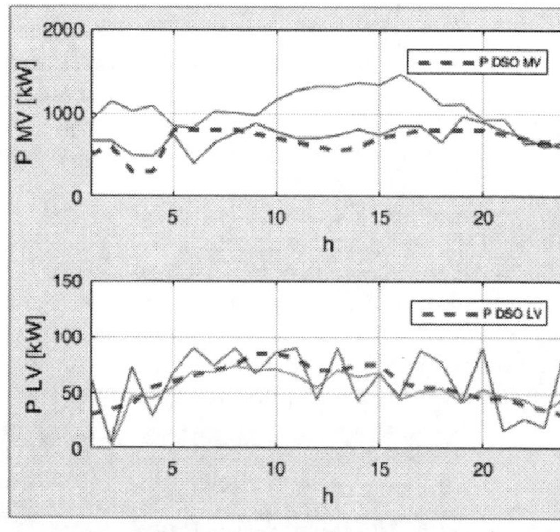

Figure 5. The graphic comparison between: MV grid DSO profile and INGRID plant power consumption for OF2 best solution; LV grid DSO profile and INGRID plant power generation for OF3 best solution.

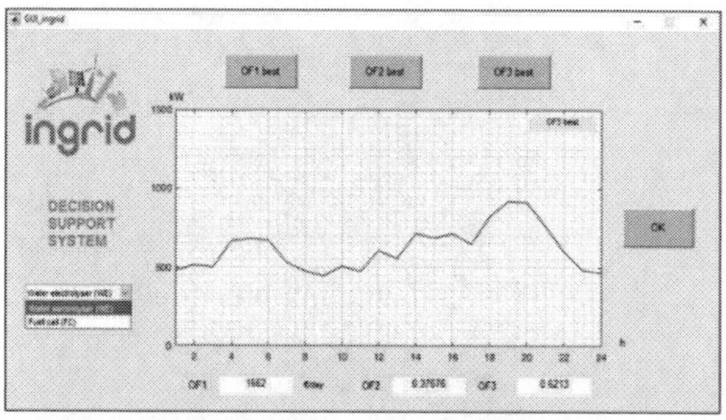

Figure 6. Graphic User Interface (GUI) of the DSS developed in MATLAB® environment.

VI. CONCLUSION AND FUTURE WORK

This paper proposes the EMS for a multi-carrier energy system that involves both electricity, as well as hydrogen and interconnects three different energy distribution networks, providing them flexibility and balancing services.

The first part of the present study addresses the structure of the proposed system and the concept of multi-carrier hub. After that, the EMS structure is explained, along with all its modules. In particular, the optimization framework and its multi-objective algorithm are shown and analyzed. This multi-objective optimization process provides very good results and allows to implement very complex management criteria driven by different objective functions. In future studies, this kind of approach can be used for the optimization tasks of other smart grid or microgrid implementations, e.g., for the management of a district level and/or a building level power flows scheduling.

The last two sections explain how the data obtained by means of the optimization process are exchanged by this application and the plant equipment and how they are employed for defining the real plant configuration, as well as for monitoring purpose. The OPC protocol has been chosen in order to ease the communication with the industrial devices. Today, other protocols, such as IEC61850 and OpenADR, are under investigation for allowing a more flexible negotiation framework between the figures asking for a service, like the DSO in this study, and the systems in charge of fulfilling them.

ACKNOWLEDGMENT

This work is part of the INGRID project, co-funded by the European Commission within the FP7 Framework Programme. Authors thank the members of the INGRID Consortium, as well as the European Commission for supporting any project dissemination activities. This work reflects only the authors' views. The Commission is not liable for any use that may be made of the information contained therein.

REFERENCES

[1] G. Andersson, F. Fröhlich, T. Krause, and A. Vaccaro, "Multiple-energy carriers: modeling of production, delivery, and consumption," Proc. 2011 IEEE, Feb. 2011, vol. 99, pp. 15-27, doi:10.1109/JPROC.2010.2083610.

[2] M. Geidl, "Integrated modeling and optimisation of multi-carrier energy system," Ph.D. dissertation, Power Systems Laboratory, Swiss Federal Institute of Technology (ETH), Zurich, 2007.

[3] A. Shabanpour-Haghighi and A. Reza Seifi, "Multi-objective operation management of a multi-carrier energy system," Energy, vol. 88, pp. 430-442, Aug. 2015, doi:10.1016/j.energy.2015.05.063.

[4] K. Kampouropoulos, F. Andrade, E. Sala, and L. Romeral, "Optimal control of energy hub systems by use of SQP algorithm and energy prediction," 40th Annual Conference on Industrial Electronics Society (IECON 2019) IEEE, Nov. 2014, pp. 221-227, doi:10.1109/IECON.2014.7048503.

[5] S. Skarvelis-Kazakos et al., "Multiple energy carrier optimisation with intelligent agents,", Applied Energy, vol. 167, pp. 323–335, Apr. 2016, doi:10.1016/j.apenergy. 2015.10.130.

[6] L. M. Ramírez-Elizondo and G. B. Paap, "Scheduling and control framework for distribution-level systems containing multiple energy carrier systems: Theoretical approach and illustrative example," International Journal of Electrical Power & Energy Systems, vol. 66, pp. 194-215, Mar. 2015, doi:10.1016/j.ijepes.2014.10.045.

[7] Z. Pan, Q. Guo, and H. Sun, "Interactions of district electricity and heating systems considering time-scale characteristics based on quasi-steady multi-energy flow," Applied Energy, vol. 167, pp. 230–243, Apr. 2016, doi:10.1016/j.apenergy. 2015.10.095.

[8] M.H. Shariatkhah, M.R. Haghifam, M. Parsa-Moghaddama, and P. Siano, "Modeling the reliability of multi-carrier energy systems considering dynamic behavior of thermal loads," Energy and Buildings, vol. 103, pp. 375-383, Sep. 2015, doi:10.1016/j.enbuild.2015.06.001.

[9] M. Moeini-Aghtaie, A. Abbaspour, M. Fotuhi-Firuzabad, and E. Hajipour, "A Decomposed Solution to Multiple-Energy Carriers Optimal Power Flow," IEEE Transactions on Power Systems, vol. 29, pp. 707-716, Mar. 2014, doi:10.1109/ TPWRS.2013.2283259.

[10] R. Niemi, P.D. Lund, and J. Mikkola, "Urban energy systems with smart multi-carrier energy networks and renewable energy generation," Renewable Energy, vol. 48, pp. 524-536, Dec. 2012, doi:10.1016/j.renene.2012.05.017.

[11] INGRID (High-Capacity Hydrogen-Based Green-Energy Storage Solutions for Grid Balancing) EU FP7 Project: http://www.ingridproject.eu/ [retrieved: January, 2016].

[12] G. Paternò et al., "Smart Multi-carrier Energy System: Optimised Energy Management and Investment Analysis," Proc. IEEE Energycon 2016, in press.

[13] SCADA System SIMATIC WinCC - HMI Software – Siemens: http://w3.siemens.com/mcms/human-machine-interface/en/ visualization-software/scada/pages/default.aspx [retrieved: January. 2016].

[14] OPC Fundation: https://opcfoundation.org/ [retrieved: January, 2016].

[15] C.A. Coello, G.B. Lamont, and D.A. Van Veldhuizen, Evolutionary algorithms for solving multi-objective problems, 2nd ed., New York, Springer, pp. 5-29/91-94, 2007.

[16] K. Deb, S. Agrawal, A Pratap, and T.A. Meyarivan, "A fast elitist non-dominated sorting genetic algorithm for multi-objective optimisation: NSGA-II," Lecture Notes in Computer Science, vol. 1917, pp. 849-858, Sep. 2000.

[17] R.P.J. Benvindo, A.M. Cossi, and J.R.S. Mantovani, "Multiobjective short-term planning of electric power distribution systems using NSGA-II," Journal of Control, Automation and Electrical Systems, vol. 24, pp 286-299, Jun. 2013, doi:10.1007/s40313-013-0022-5.

[18] B. Tomoiaga, M. Chindris, A. Sumper, A. Sudria-Andreu, and R. Villafafila-Robles, "Pareto optimal reconfiguration of power distribution systems using a Genetic algorithm based on NSGA-II," Energies 6, pp. 1439-1455, Mar. 2013, doi:10.3390/en6031439.

[19] D. Buoro, M. Casisi, A. De Nardi, P. Pinamonti, and P. Reini, "Multicriteria optimisation of a distributed energy supply system for an industrial area," Energy, vol. 53, pp. 128-137, Sep. 2013, doi:10.1016/j.energy.2012.12.003.

[20] D. Arnone et al., "Mixed heuristic-nonlinear optimisation of energy management for hydrogen storage-based multi carrier hub," Proc. IEEE International Energy Conference (ENERGYCON 2014), May 20144, pp. 1019-1026, doi: 10.1109/ENERGYCON.2014.6850550.

[21] G. Paternò et al., "A matrix model for an energy management system based on multi-carrier energy hub approach," Energy 2015 IARIA, May 2015, pp. 18-23, ISBN:978-1-61208-406-0.

International Academy, Research, and Industry Association (IARIA)
412 Derby Way
Wilmington, DE 19810

ISBN 978-1-5108-2580-2

9 781510 825802

CURRAN ASSOCIATES INC.
proceedings.com

Eleventh International Conference on Advances in System Testing and Validation Lifecycle (VALID 2019)

Held at SoftNet 2019

Valencia, Spain
24 – 28 November 2019

IARIA